FLYWHEELS

HOW CITIES ARE CREATING THEIR OWN FUTURES

FLYWHEELS

TOM ALBERG

Columbia Business School
Publishing

publication supported by a grant from
The Community Foundation *for* Greater New Haven
as part of the Urban Haven Project

Columbia University Press
Publishers Since 1893
New York   Chichester, West Sussex
cup.columbia.edu

Library of Congress Cataloging-in-Publication Data
Names: Alberg, Tom, author.
Title: Flywheels : how cities are creating their own futures /
Tom Alberg, Columbia Business School Publishing.
Description: New York : Columbia University Press, [2021] | Includes index.
Identifiers: LCCN 2021017376 (print) | LCCN 2021017377 (ebook) |
ISBN 9780231199544 (hardback) | ISBN 9780231553186 (ebook)
Subjects: LCSH: Cities and towns—Effect of technological innovations on—
United States. | Urban renewal—United States. | Technology—United States—
Sociological aspects.
Classification: LCC HT167 .A5738 2021 (print) | LCC HT167 (ebook) |
DDC 307.760973—dc23
LC record available at https://lccn.loc.gov/2021017376
LC ebook record available at https://lccn.loc.gov/2021017377

Cover design: Noah Arlow

To entrepreneurs, inventors, and all
who are working to build a better world.

CONTENTS

PART 3

PART 4

FOREWORD

I first met Tom Alberg in 1987 when I was trying to purchase shares in a small, struggling tech company. Tom was the managing partner of Perkins Coie, the largest law firm in Seattle. He had done work for the tech company, and when it could not pay its bills, Tom agreed to take stock instead. Over lunch in Kirkland, I got to know this curious, fiercely intelligent but gentle man, who had stewarded Boeing and several other Seattle corporations through various legal challenges. Tom told me that he liked small companies, and he thought Perkins Coie would hang on to its stock. His insights in that early stage company were prescient, and Tom went on to found the largest venture capital firm in the Pacific Northwest, demonstrating similarly keen instincts for the next three-plus decades.

In 1990, Tom left the practice of law to join McCaw Cellular, where I was the vice-chair. As Tom recounts in the book, we met to talk about his compensation package, which included stock options, again demonstrating Tom's belief in betting on the future. Since that time we have served together on several boards, both public and nonprofit. More recently my Trilogy Equity partnership has invested in nine businesses with Tom's venture capital group, Madrona, and I have personally invested in every Madrona fund. It is not too much to say that Tom, still providing leadership and guidance to enthusiastic young entrepreneurs, is at eighty, an éminence grise within his firm, the wider Seattle business community, and indeed among tech's venture capitalists at large.

That is why I was so excited to read his take on the remarkable story of Seattle's tech explosion and to add my thoughts. A native of the city and longtime investor myself, I've had the great fortune to walk with Tom along this journey.

In 1970, shortly after Tom returned to Seattle from his early career in New York, the city was a blue-collar community anchored in manufacturing, but its largest employers—Boeing, Paccar, and Weyerhaeuser—were in sharp decline. Fifty years later, the city is home to two of the three largest companies in the world, Amazon and Microsoft. And the next three largest companies in the Puget Sound region—Starbucks, T-Mobile, and Costco—are all larger than Boeing. Those five hometown success stories were all started by entrepreneurs, three of whom grew up in the Seattle area.

Tom's book tells the story of that transformation and postulates that there is a growth "flywheel" by which momentum is created within a community to build new businesses. The presence of 140 other software companies in the Seattle area, including offices of every major global tech company, evidences that Seattle's growth flywheel has been a transformative success.

The author tells the story of this evolution in Seattle's business community and describes the development of the growth flywheel. In many ways, Tom's career parallels Seattle's journey. After returning home to the city from New York in 1967, Tom would become the principal outside counsel at Boeing and lead Perkins Coie to become Seattle's preeminent law firm. Always nimble, he transitioned to help lead the cellular revolution at McCaw Cellular, and then created the city's dominant venture capital firm, providing financing for more than two hundred startups. While Amazon employs "Day 1" as a philosophy, Tom was a Day 1 investor in the company and served on Amazon's board of directors for a quarter-century.

Seattle's economic renaissance was the consequence of both necessity and leadership. The late 1960s were marked by the near failure of Boeing, where employment declined from more than one hundred thousand in 1969 to under forty thousand just two years later. The despair and resulting migration out of Seattle were captured on a now-famous billboard, erected in April 1971, that read, "Will the last person leaving Seattle—turn out the lights."

Despair turned to design. Community leadership came not only from political leaders but also that same bright, young lawyer, Tom Alberg.

The group known as Forward Thrust passed an ambitious set of ballot initiatives for sewers, fire protection, parks, roads, neighborhood improvements, and the first professional sports stadium (the Kingdome). But critically, people failed to pass a proposed rapid transit system, community centers, and low-cost housing programs. Educational leadership came from the University of Washington, thanks in part to the commitment of federal funds. Ultimately, the UW would become a hub for computer science and biomedical innovations that have spawned dozens of companies, thanks to funding and commitments made in the 1960s.

The book describes two flywheels: a business growth flywheel that has worked miracles; and a second system that Tom describes as a livability flywheel. This livability flywheel has taken up ever-greater importance in his vision of the future—of Seattle and all cities that experience comparably explosive growth. Despite Seattle's astounding economic growth, Tom points out its painful failings in public education, transportation, and homelessness.

During the fifty-year transition chronicled in this book, Seattle went from an economy based largely on manufacturing jobs to STEM-based employment that draws hundreds of thousands of computer science graduates from across the country and around the world. But investment in public education has not kept up. Gridlock is a constant on our roads. And while Seattle's population has grown by more than one-third since 2000, housing starts have not kept pace, resulting in rampant gentrification and a terrifying rise in homelessness, which now places Seattle with the third-highest homeless rate in the United States. The three-headed challenge of 2020—a global pandemic, resulting economic disruption, and exploding social justice crises—challenged Seattle anew. Tom, ever the community leader, has worked to address many of these problems, serving on countless commissions, committees, and task forces. An eternal optimist, he can find glimmers of progress in each area, and now he implores readers as he has implored business and political leaders to come together around solutions.

Tom proposes that the more challenging flywheel is the livability flywheel because, while the Seattle area has had stunning economic growth, many of its citizens have been left behind and the political culture has been paralyzed by a backlash, or "techlash," against those companies and entrepreneurs that built the new capitalist pillars of the community. As a board member at Amazon, Tom has been on the front line, watching the e-commerce giant become a lightning rod of problems that can accompany success. Through it all, he has continued to address progress and challenges in education, transportation, housing, and environmental stewardship. After all, Seattle is his hometown.

This vantage point, as both native son and new-economy leader who has participated in virtually every aspect of the city's development, uniquely positions Tom Alberg to comment on the growth of Seattle over the last half-century. As a lawyer, executive, board member, investor, and, crucially, as a social entrepreneur involved in everything from autonomous vehicles to organic farming and the challenges of homelessness, Tom has been one of the seams holding the quilt of the Pacific NW together.

The question persists: Why have Seattle and San Francisco dominated the tech economy, rather than cities like Chicago, LA, Detroit, or Philadelphia? Tom offers his theories. Each of the other cities had advantages of population, educational institutions, geography, and history that should have positioned them to thrive, particularly in comparison to Seattle, which was far smaller in 1970 and widely viewed as Boeing's company town. Seattle's ascent appears to prove Tom's point that "any city can create an economic flywheel tailored to its particular strengths." In chapter 7, he shows us similar flywheels beginning to gain traction in Oklahoma City, Tulsa, and Kansas City, bringing readers along to understand how those cities are striving to create flywheels that balance economic growth with livability. No doubt, Tom is trying at this very minute to bring home lessons learned from these Midwest successes.

This book is ultimately about leadership successes and failures. While Tom has succeeded as an investor, he excels at creating culture. I've seen that firsthand in our work together at McCaw Cellular and in the community. He is often asked which is more important to build a successful company, a great plan or a great leader. Though judicious in manner, Tom's innate curiosity and imagination always shine through. "Both," he says with a wry smile, perhaps because he is himself a great leader with a great eye for spotting the technological innovations that can change a city's economy and future.

John Stanton
Managing partner, Trilogy Equity Partners, and chairman, Seattle Mariners
Member of the board of directors of Microsoft and Costco
Former CEO of VoiceStream Wireless, renamed T-Mobile

FLYWHEELS

Part 1

Prelude to Jeff Bezos's Day 1

This is the story of how Seattle, once a sleepy lumber and shipping outpost in the far northwest corner of our country, became one of the two leading tech centers in the world, along with Silicon Valley, and how business, non-profit, and government leaders in other cities can build their tech-driven economies and create prosperous cities. Although many cities across the United States are already planting the seeds for growth, they need to accelerate and broaden their efforts to be successful.

A tech-driven economy is not sufficient, however, to meet the many challenges that confront twenty-first-century American cities. We also must figure out how to solve the civic and social problems of our big cities: mounting homelessness, fraying public safety, and increasing traffic congestion. Also unsolved are the older, deeper problems of systemic discrimination and inadequate public education, particularly for children of color. My central thesis is that we can best solve these intertwined problems when city governments and tech companies work together on solutions—in ways that go beyond lip service and advisory committees. Otherwise, the economic success and livability of our cities are jeopardized.

At the end of 2019, the country celebrated the lowest unemployment in fifty years, at 3.5 percent. Seattle and San Francisco were booming with a growing class of creative workers attracted by tech jobs, city amenities, and opportunities for interacting with friends and colleagues. The downtown streets were filled with shoppers, and restaurants and bars overflowed. Amazon and Microsoft, both headquartered in greater Seattle, were two of the three most valuable companies in the world and were acclaimed as the second- and sixth-most admired companies in 2019 (and second and third by 2021).[1] Over 140 tech companies from Silicon Valley and elsewhere had established software engineering offices in the Seattle area, some with over five thousand employees. The vibrant startup community included hundreds of cutting-edge tech companies. Artificial intelligence, biotech, quantum computing, the internet, and wireless communications promised an even brighter future. Amazing progress was being made in understanding and manipulating our genetic code. We were beginning to use gene splicing and immunotherapy to cure what were previously believed to be incurable cancers and diseases.

But all was not as positive as appeared on the surface.

In late February 2020, I pulled on the blazer I keep in my office for formal meetings (eschewing a tie, a longstanding tech tradition) and walked four blocks to an intimate dinner meeting of the CEO members of Challenge Seattle in the elegant top-floor office suite of a downtown office building. CEOs of Microsoft, Boeing, Nordstrom, Costco, Alaska Air, and eight other major Seattle companies were present. The members of Challenge Seattle are eighteen CEOs of the leading businesses and nonprofits in Seattle (I'm a member even though not a CEO, as I am the de-facto representative of the private tech companies locally.) Challenge Seattle will reappear later in these pages, as I discuss the ways that business can and should contribute to alleviating civic problems. But on this night, we were about to confront something that would upend our lives.

Midway through our meal, Susan Mullaney, northwest president of Kaiser Permanente Healthcare, interrupted our otherwise congenial discussion with a grave warning: we should prepare for a global pandemic. Steve Davis, then CEO of PATH, the global healthcare organization, reinforced her remarks. I remember thinking that if the two most knowledgeable healthcare executives at our meeting were worried, I should be worried. I promptly sent an email to my fellow partners at Madrona Venture Group that we should take the virus seriously. I think the others present departed the meeting as worried as I was—but I don't know that any of us fully anticipated the breadth of the virus's impact on everything from our economy to daily habits well into the future.

Four days later, on February 29, the first death from COVID-19 in the United States was reported in Kirkland, a Seattle suburb. The bad news traveled fast through Seattle's business and political leaders, and Seattle was later praised in the East Coast press for reacting quicker than some other major cities, including New York City.[2] Despite our early warning, as late as March 7, the Seattle Sounders were allowed by local public officials to play a soccer match before thirty-three thousand fans, even though by then there were ten known deaths and sixty confirmed cases in our area.[3] Microsoft, Amazon, and other local companies had begun encouraging their employees as early as March 3 to work from home, but it was not until March 23 that Governor Inslee, who had just ended his presidential bid, issued his "Stay Home, Stay Healthy" proclamation.

The virus showed us that notwithstanding all of our technological prowess, we were unable to prevent or quickly contain a virus that sent millions of people to hospitals, resulting in a staggering number of deaths. The virus forced us to wear masks; closed schools, businesses, and restaurants; and crippled our economy and civic institutions. It also emptied our downtowns as offices, stores, restaurants, and bars closed.

On May 25, George Floyd, a Black man was killed by white police officers during an arrest in Minneapolis. This set off worldwide protests in hundreds of cities. The majority were peaceful, but some turned into smashed windows, fires, and looting by individuals taking advantage of the unrest. Our normally bustling downtown streets already deserted by COVID-19 became canyons of concrete and boarded-up storefronts. The homeless began to pitch tents on sidewalks and in parks. Well-to-do city residents fled to their vacation houses. Others stayed away. Black Lives Matter exposed the racism in our society and our failure to provide adequate economic opportunity and education for our minority populations. Astonishingly, 67 percent of Black sixth graders in Seattle are unable to read at grade level compared to 18 percent for whites. Similarly, for math abilities.[4] And they never catch up.

What are the long-term implications of the protests and COVID-19 for our society? In this book, I make predictions but, more importantly, I advocate specific changes, policies, and programs that we should pursue to improve our lives. For example, I propose specific actions to increase the participation of Blacks and other minorities in the technology and venture worlds. I also address the urgent need to improve the quality of education at all levels for minorities and low-income students so they can be equal participants in our society.

We also should learn from our forced use of Zoom and other communications technologies to combine the best of remote working, education, and healthcare with the benefits of physical interactions. Not every meeting or healthcare appointment needs to be in-person, but we also need physical interactions. Workers do not need to be in the office every day. Schools should learn from techniques that were used successfully by some teachers and schools during the epidemic and adopt a hybrid of in-person and remote learning and communication.

We'll come back to these issues later in this book.

∼

In February 1967, I left New York City, where I had been a young associate at the law firm of Cravath Swaine & Moore, and came home to Seattle. New York City bubbled with excitement for a young person as the center of finance, law, and publishing. Cravath was known for representing some of America's great inventors, including Morse, Goodyear, Westinghouse, and Tesla (Nikola, the inventor, not the auto).

I was raised in Seattle, and it was where my family had lived for generations. When I returned, Seattle was a lot smaller and quieter than New York City and what Seattle is today. It was mostly a one-company town, dominated by the Boeing Company, its biggest employer by far. Boeing rolled out the 737 that year and would launch the 747 three years later. Computers were mainframes that filled entire rooms. Windows were made of glass. Clouds were plentiful but were decades away from revolutionizing the way we use computers. The word *biotechnology* was hardly heard of.

Seattle was solidly middle class. Its public university was good, but nothing like the research powerhouse it is today. In addition to Boeing, the leading businesses headquartered in the Seattle area represented a cross section of traditional companies: retailing (Nordstrom), truck manufacturing (PACCAR), timber (Weyerhaeuser), hospitality (Western Hotels), electric distribution (Puget Sound Power & Light Company), and medium-sized local banks (Seattle First Bank, Rainier Bank, and Peoples Bank). Major players in the business community also included the two Seattle daily newspapers, lawyers, contractors, and developers. Microsoft and Amazon were in the future. Leading businessmen and lawyers decided most civic issues over cocktails and lunch at the exclusive Rainier Club downtown, where dominoes was the favorite game. Local public officials were cautious and parochial, with a tinge of corruption because of a police tolerance policy on gambling and other vices. But Seattle was not a wide-open town—it could best be characterized as sleepy. After all, that cloudy, rainy weather is conducive for a good night's rest.

Half a century later, the city is fully caffeinated. It has a diversified tech economy, anchored by Microsoft, Amazon, T-Mobile, Costco, and Starbucks, and bolstered by hundreds of startups headquartered in greater Seattle. In numerous technology categories—like cloud computing, machine learning, artificial intelligence, e-commerce, the commercial exploration of space, the development of autonomous vehicles, and virtual and augmented reality—Seattle and Silicon Valley are the superpowers. Attracted by our growing pool of engineering talent, the 140 global tech firms that have established software engineering offices here range from Google, Facebook, and Apple to Alibaba, SpaceX, and Blue Origin.

One way or another, I've worked for or brushed shoulders with most of our leading companies. I've represented, advised, and invested in Boeing, McCaw Cellular, Amazon, Immunex, Alaska Airlines, and many smaller tech companies. I participated in the birth of the mobile phone revolution as executive vice president at McCaw Cellular and then president and chief operating officer of its 50 percent publicly owned LIN Broadcasting, where I was responsible for the cellular operations in New York City, Dallas, Houston, and Los Angeles. The venture capital group I cofounded in 1995, Madrona Venture Group, has invested over $1.5 billion in more than two hundred startups in the Pacific Northwest.

But when reporters tell my story, most of them boil this success down to two key events: my first meeting with Jeff Bezos in April 1995 and my decision a few months later to become one of the first investors in his internet bookstore idea, Amazon.com, where I served as the longest-sitting director other than Jeff himself.

Accordingly, this book is also about Amazon and Jeff and their enormous impact on Seattle, the tech world, and customers across the globe. Now with Jeff's February 2021 announcement that he is stepping down as Amazon's CEO, we will witness his second act, devoted, I predict, to helping conquer outer space with Blue Origin and solving some of our most pressing problems, including global warming, education, and homelessness. He has been planting many of these seeds in recent years, and we should all expect that twenty years from now, he will be as well-known for these accomplishments as he is for Amazon.

The groundwork for my first meeting with Jeff had been laid a couple of weeks before in the spring of 1995 when I got a phone call from a lawyer named Tom Foster. Tom was a partner at Foster Pepper, a small but highly respected law firm where he represented a lot of Seattle's movers and shakers in the developer world. I had met Tom when I was a partner at Perkins Coie, and we had worked on several business deals over the years. We'd become friends. Every summer I spent several days anchored on my sailboat in the harbor of Mink Island, which he owned with a couple of his clients, in Desolation Sound, British Columbia.

Tom belonged to a small investment group of his developer clients. Through one of its member's young adult children, the group had heard about this guy Jeff Bezos, who was trying to raise money for an internet startup. The investment group wanted to know if it made any sense as an investment. "You're at McCaw," Tom said to me in a telephone call, "so you must know a lot about the internet and tech stuff like that. Could you talk to him?"

A former hedge fund analyst, Jeff had moved from New York to Seattle the year before, specifically planning to create an internet bookstore. It was a heady time in the city. Seattle was in the headlines from grunge music culture to technology.[5] Microsoft released Windows 95 that fall, and AT&T completed its purchases of Seattle's McCaw Cellular and LIN Broadcasting companies for a combined $18 billion (deals in which I played key roles), giving the communications giant control of a national mobile phone network. Jeff and his wife MacKenzie launched their Amazon website on July 16, 1995.

I was hardly an expert on the internet, but then as now I always made time to learn about new technologies or businesses, even if there is no immediately apparent opportunity for me. So, despite being mired in negotiations with AT&T, demanding dozens of conference calls and trips to New York, I told Tom, "Sure."

I did know a bit more about the internet's commercial potential than most people at the time. My twenty-four-year-old son John was preparing to launch one of the first business-to-business (B2B) web services companies.[6] Several of the executives with whom I'd worked closely at McCaw had just departed to become the CEO, CFO, and general counsel of Netscape, the first commercial internet browser. Coincidentally, I had passed up an opportunity to get in on the ground floor of the internet two years earlier when I had taken several McCaw executives to a meeting with a pair of Seattle techies at a small, local engineering company that had created a crude graphical user interface that could be used to view information from the internet. Today I recognize their invention was an early browser like what Netscape, Chrome, and Foxfire became.

Unfortunately, the executives and I failed to understand the interface's full significance, and when the techies didn't attract any investors, they never pursued their idea.

In April, a few days after I spoke with Tom, Jeff called and introduced himself. We set up a time for a meeting, and he said he would send me his business plan in advance. I've still got it in my office. It was a photocopy of an original that he'd given to somebody else—their name was still on its cover. He'd scribbled a hand-written note to me: "This is a bit out of date now, but it does a good job describing our vision. Jeff." Forty-seven single-spaced pages, it describes in rich detail the online internet bookstore he envisioned. It includes succinct descriptions of the internet; thoughtful analysis of book pricing and the publishing industry at large; and long descriptions of the marketing, advertising, and public relations initiatives the business would deploy. Its bottom-line summary was: "More titles, lower prices."

Amazon's site was not yet live, but Jeff had been developing it for about a year at that point, using his own money—and his parents'—to pay the bills. Believing that software developers "are better characterized as artisans or craftsmen than as scientists or engineers" and that "the productivity of a programmer . . . can range over a factor of one hundred or more," he had carefully selected two software developers to write the code.[7] MacKenzie, Jeff's wife, was in charge of finances. They had recently moved the operation from their garage into a warehouse south of downtown Seattle where the rent was $995 a month, including utilities—an increase from the $293.70 they'd charged themselves for the garage.

Several of the major themes that Jeff sounded back then remain core to Amazon's philosophy today, including the fundamental importance of software developers, rigorous hiring standards, and the competitive advantage in offering customers low prices, convenience, and broad selection. At the time, his forecasts sounded wildly bold. "The company expects to take its stock public in 5 years, when it expects its annual sales to be $65 million, its pre-tax income to be $4.6 million, and its valuation to be in the range of $90 million to $130 million,"[8] he'd written. Those projections turned out to be wide of the mark. In 2000, five years after we met, Amazon's annual sales were not $65 million but $2.7 billion.

This extraordinary growth was offset by losses of $1.4 billion—rather than breakeven. At the end of 2000, Amazon's stock value was $5 billion, thirty-eight times Jeff's high-end prediction of $130 million.

We met for the first time several days later at my LIN Broadcasting office in Kirkland, across Lake Washington from downtown Seattle. Jeff struck me as very smart and confident but not overbearing.[9] He was passionate about his idea, but his arguments for it were well reasoned. He was engaging. And he had that great, deafening laugh that has become a hallmark. Jeff didn't take me through the document systematically, as I'd assumed he would. Instead, he talked off the cuff about what he was trying to build and why. He told me how he'd become intrigued with the internet while working as an analyst at the hedge fund D. E. Shaw in New York; though the technology was still in its infancy, he'd come to believe that someday soon it would connect hundreds of millions of people, all around the world.

Although the internet's origins date back to the late 1960s when researchers developed the ARPANET so the Department of Defense computers could exchange messages, use of the internet did not begin to grow significantly until 1990, when Tim Berners-Lee developed the first web browser, effectively inventing the World Wide Web. Usage spread quickly when programmer Marc Andreessen launched his Mosaic browser (later Netscape) in 1993, which made it possible to include color graphics and pictures.

Jeff was inspired to start Amazon in 1994 after he read a study that said internet usage was growing by an annual rate of 2,300 percent. For Jeff, this was "Day 1" of a future of unlimited opportunities—a theme we shall see that Jeff carries to this day that we are still in Day 1.

Hardly any internet companies that we would recognize today existed at that time, although eBay was launched in 1994 as an auction venue for third parties selling used items and Yahoo began as a site that listed interesting websites. A few sites sold music and items such as flowers to retail customers. There already were a couple of online booksellers. The use of email was just beginning, spurred by the rapid growth of AOL, which signed up five million subscribers by 1996 after flooding the country with mass mailings of its installation CDs.

Convinced that the business opportunities of the web had hardly been scratched, Jeff researched several different ideas and settled on book retailing, because the internet could provide a complete and easily searchable inventory of virtually every book in existence—something a physical store could never match. Brick-and-mortar stores were limited not only by space constraints but by how much money they were willing to tie up in unsold inventory. All Jeff needed to do was list books that were in print; when a customer bought one, Amazon simply ordered it from its publisher or a wholesaler. If he ran the operation efficiently, he believed he could deliver most titles in just two or three days and make a profit.

We discussed pros and cons, such as the presence of the giant Barnes & Noble in the book market. It wasn't brilliantly obvious—at least not to me—that Amazon would be as big as it is today. It was an intriguing idea, though, and I could see the potential.

Jeff had originally incorporated the company as Cadabra, a reference to the magic of the internet, but his lawyer suggested to him that it sounded too much like *cadaver*. The story is that he named his store Amazon after the river because it is so large—but probably equally important because having a company whose name began with the letter *A* meant that it would always be at or near the top of any list of internet retailers. (Lists were an important way for users to find sites in those early days—though there were a few search engines, Google didn't publicly launch as a search engine until 1998.)

Jeff told me that he had spent several months raising roughly half of the $1 million that he needed to get his business off the ground. I told him that it sounded like he was going door-to-door to ask for it. What I didn't know was that much of this seed money had come from his parents.

I wish I'd been prescient enough that I could tell people today, "Oh, I could see how brilliant Jeff was," or that I immediately recognized that Amazon would get to be a hundred-billion-dollar company. But the truth is I couldn't have imagined the full extent of what Amazon would become, and I don't think Jeff did either. But he did not doubt that he could compete with bigger companies.

"What about Barnes & Noble?" I asked him.

Remember that by the mid-1990s Barnes & Noble had become the dominant bookseller in America. In the same way Walmart was seen as threatening Main Street retailers in the 1980s, Barnes & Noble was viewed as the enemy of small independent bookstores. The 1998 film *You've Got Mail* starring Tom Hanks as the owner of a Borders or Barnes & Noble look-alike was pitted against Meg Ryan, the owner of a boutique bookstore. Jeff's online bookstore was laughable within that context, but he was confident.

"I think I can move faster than them. And I think they probably won't compete aggressively online because it would cannibalize their huge investment in physical stores."

"Well, I love going into bookstores," I said. "It's nice to feel the books. It's sort of a social experience."

Jeff replied, "I like bookstores, too, but they don't have the inventory. And there are other disadvantages. But there's room for both."

We talked about his love of books, and I was to see ample evidence of it in the coming years. Jeff has always liked what he calls "long-form content," which might surprise some people since it's the antithesis of much of what's popular on the internet. He regularly holds book club sessions with his top executives and sometimes with the Amazon board. You can learn a lot about his thinking from his syllabus. The first book he assigned us, *The Innovator's Dilemma* (1997), was by Clayton Christensen, who conceived the disruptive innovation theory of business. Jeff used it to emphasize a historical truth that guides his thinking: nearly all of the most successful companies of the last hundred years eventually succumbed to younger challengers, who innovated with new products or business models, while the established company failed to change its strategies because they were so lucrative. The only way to meet such challenges, Jeff believes, is to aggressively pursue new strategies, even at the expense of undercutting your core business in the short term. Another book for our book club was Jacquie McNish's and Sean Silcoff's *Losing the Signal* (2015), which describes the demise of BlackBerry.

Jeff and I discussed the potential investment by Tom Foster and his group, but he also pitched me, directly. "I need to think about that," I said. "I'm intrigued, especially by the way you want to use the internet. So let me think it over more. In the meantime, I'll talk to Tom."

According to the financials Jeff gave me a few months later, breakeven would occur in September 1996 when cash flow would be positive, with monthly revenues of $707,019. Jeff said it would be easy to keep the operating expenses low because the company didn't hold any inventory. Jeff would fax a list of the day's orders to a large wholesale book distributor called Ingram that had a warehouse in Roseburg, Oregon, and they would send the books by UPS for delivery to Amazon in Seattle the next day. At the Amazon offices, Jeff and MacKenzie helped repack the books for shipment to individual customers, who received them in just a few days, depending on how far they were from Seattle. Today, of course, Amazon stocks millions of items in its fulfillment centers spread across the United States as well as in a growing number of other countries, and it can deliver many of them within an hour.

Amazon's early incarnation had another advantage, which continues to this day: customers paid by credit card. That meant instant cash for

Amazon, while its payment terms with suppliers are in thirty days or more. The bigger they got, the greater their cash flow advantage.

After Jeff left, I called Tom and said, "I met with Bezos. He's got a good plan, and he's for real. So is the internet. You guys make your own decision. But I might well invest." It just wasn't yet entirely clear to me that Amazon *was* a fabulous investment, but it didn't need to be entirely clear. Tom thanked me for confirming for him that this was not a crazy idea, at least not from the technology side. He would talk to his fellow investors.

A couple of weeks later, in early May, Tom called me back. "We talked to Jeff, and we think his valuation is too high"—Jeff's postmoney valuation of Amazon was around $5.2 million—"we told him that we would invest if he lowered it to $4 million. But Jeff said no, he was going to stick with his valuation. So we're not going to invest."

I replied, "Well, I'm not that sensitive to whether his valuation is $5 million or $4 million. I might invest."

By then I was feeling quite positive about Jeff's proposal. After all, it was the *internet*, which was already growing exponentially. I felt that I could probably invest $50,000. But I still needed to think about it. I'd made angel investments of $10,000 and $20,000 in other tech companies. But $50,000 was a lot of money for me at that point; in fact, it would be the largest investment I'd ever made.

A couple of weeks later, in June, Jeff called me again. I told him, "I'm really pretty interested, but I'm not quite ready to make a decision. Let's wait and see when you launch."

The truth was, I had mostly made up my mind. One of the things that swayed me was a visit to a local bookstore to buy a gift for my son John. I had a specific book in mind for him, because I thought he would find it helpful while he was setting up his own internet business. I couldn't quite remember the title, though I did remember that the author was Peter Drucker. I couldn't find it on the shelves myself, so I asked a clerk for help. He searched around, and at first he couldn't find the author. Then when he found the shelf that had the Drucker books, the one I wanted, *The Effective Executive* (1966), wasn't in stock. (Much later, I learned that this is one of Jeff's favorite books and that he recommends it to his managers.) As I set out for another store, where I hoped to have better luck, I thought, "You know, it would have been great to *go on the internet* and buy the book. I wouldn't have to drive all over town. It would just come to me."

I should mention that today I shop online and in person, frequenting Seattle's many quality bookshops that cater to local tastes and have built convivial communities.

When Amazon launched on July 16, 1995, its home page looked nothing like it does today. It was black and white, with clickable words or phrases in blue. There were no fancy graphics (unless you count a somewhat stylized large *A* in a blue box), and it contained only a few images of book covers. The book descriptions were limited, dry, and I complained to Jeff that they lacked publication dates—essential information if one wanted to buy the latest edition of a travel guide, for example. But the site listed hundreds of thousands of books in its inventory, any of which could be delivered in five days or less.

Over the next few months, I would periodically get phone calls and sometimes a fax from Jeff. "Well, we're making progress," he'd say. "We've sold books in five states now." One time he called and said, "We've got our first European order!" He sent me weekly sales figures. For example, for the week beginning July 23, 1995, his second in business, he'd received orders totaling $14,792. Sales continued at roughly this rate, so in August I called him back.

"Okay, I'm in for fifty thousand. But only when you've got the whole million dollars committed—I don't want to invest until then. So stay in touch and let me know when you've done it."

I had learned over the years that startups often fail not because the idea is bad but because they can't raise the necessary capital. Jeff was not put off by what you could call my careful approach. As recently as a couple of years ago, I was with him and a group of Amazon directors and executives and Jeff said, "One thing I always admired about Tom—he wouldn't put his money in until he knew I could raise all I said I needed."

Even with his parents' help, it would take Jeff a full year to raise his million dollars, most of it in chunks of $50,000 or less. Besides Foster's investment group, at least one venture fund turned him down. They told me they didn't think Amazon would be able to compete when Barnes & Noble launched its website. Over the years, people in Seattle would tell how they turned Jeff down. Foster remained a friend and loved to tell his Amazon story.

While Jeff continued to call on investors, I was working to close the sale of LIN Broadcasting to AT&T. We had closed the sale to them of McCaw Cellular the previous September. One year later, in 1995, we completed the sale of LIN and I promptly signed my termination agreement and moved out of my office and into the personal suite of Bill Gates's investment group, BGI. The office was close to Craig's in the same Carillon Point development in Kirkland. It was a nice corner office, but it didn't have a view of Lake Washington like my old office. And although I was surrounded by Bill's financial wizards, who were investing his personal wealth and Microsoft's surplus cash, I wasn't working directly for him.

Craig had called me in August to ask if I would set up a company to bid in the FCC spectrum auctions scheduled for later that fall. There had been a set of spectrum auctions in the spring, and many of the big telecom companies, including AT&T, had purchased licenses. But the FCC wanted to give smaller businesses an opportunity to acquire licenses too, so they'd set aside a very large block of spectrum for cellular service and divided it among the 493 Basic Trading Areas (BTAs) in the United States to be auctioned off to individuals and small businesses with a net worth under $50 million.

Craig assumed that I would qualify, which I certainly did. He said the FCC rules permitted others to invest in the company that I would set up, so long as I controlled it and owned 51 percent. Craig's lawyers had crafted a structure whereby Craig and a group of investors that he was talking with, including Gates, George Soros, and Saudi Prince Alwaleed bin Talaland, would pledge several hundred million dollars and pay the $40 million deposit. The FCC had adopted very liberal financing rules; if we won licenses, we would only have to pay interest for the first five years.

Trusting Craig and being adventuresome, I said yes. I immediately had to hire a team and lawyers to set up a software system to bid on the licenses. This is not the place to tell the whole story, but daily bidding began in December, and by February we were the leading contender in numerous markets—including New York City, San Francisco, and Los Angeles. The prices, however, were getting extremely high, and when our total bids reached $300 million, I called Craig and told him that I was going to stop bidding. We had avoided conferring since the start of the auctions, so as not to violate the FCC rules. His only comment was that he had watched the escalation, too, and was hoping I'd call. (As it turned out, we were prescient. Several of the winning bidders later went through bankruptcies and years of litigation because of their overpayments.) It had been an interesting distraction, but I was ready now to turn full time toward working with my partners in our new Madrona investment group.

Back in late November, when I was just plunging into the licensing auction, Jeff had called to say, "I have my million dollars committed now.

Are you ready?" In early December 1995, about eight months after first meeting him, I wrote him my check.

People sometimes ask me, "Why invest in this guy you'd only just met, and this internet business idea you'd only just heard about?" For one thing, I'd been looking for ways to invest in more startups, and I didn't own any internet companies. But more than that, it was the combination of the founder, the technology, and the business plan. I liked the idea of investing in the internet, I liked what I was hearing from Jeff, and I also liked Jeff. Those are the three things you need in any investment: (1) a talented person to launch the business; (2) a technology that itself drives growth potential, often one that disrupts convention; and (3) the business has a way to actually make money, usually with a narrowly focused starting point.

But beneath all of that was my comfort with a certain amount of risk. In this case, $50,000. I'd decided to officially stop working for large businesses to become an investor full time. My wife and I were going to live off what I'd made from the McCaw sale to AT&T and any future investments. By December, when I wrote the check to Jeff, I was already talking to the well-known public official and corporate executive William Ruckelshaus, as well as Burlington Northern Railroad CEO Gerald Grinstein, and my old law colleague Paul Goodrich about forming an investment firm. Our vision wasn't very well defined. We figured we would share an office and work together, but we would each make individual investment decisions. We weren't really forming a fund at that point, and we weren't going to collect money from investors. We were each going to invest our own money.

While $50,000 was indeed a large amount, it wasn't $1 million, and it wasn't a crazy speculation—at least in my mind. I thought investing in Amazon was a measured risk in a technology that appeared to have great potential to change lives. Everyone, however, has a different risk tolerance. When I was a lawyer at Perkins Coie, I had clients who came in to talk about the legal aspects of an investment, and when I'd mention the investment itself—saying something like, "You know, there are risks in these things, you could lose all your investment"—they'd say they "Really?" and opt out. They never considered the degree of the risk. They didn't assess whether it was relatively high or relatively low or how much uncertainty they were willing to tolerate. Later in the book I'll describe how I applied those considerations to other investment decisions, and also why it is okay if some investments don't turn out.

When I became an angel investor in Amazon, I was not just investing in Jeff's book idea. It was really about the notion that people would do many things on the internet. Jeff was entirely focused at that time on becoming the world's largest retailer of books. Only in footnote 21 of his business plan did he mention any other items: "The Company may also carry products such as pre-recorded audiocassettes, music CDs, CD-ROMS, and pre-recorded videotapes." Though he fully believed his idea would make money and grow rapidly, it took off and grew faster than even he'd predicted. And it led him into byways, like e-readers, original entertainment, web services, and the cloud, that none of us could have imagined.

Jeff wasn't ready to have anyone on his board of directors other than himself and his wife, MacKenzie. But he did want an advisory board and, to my surprise, he invited me to be one of three advisory directors. Nick Hanauer, an early investor, who worked in his family's bedding business (and has since become a well-known angel investor in Seattle), was another. The third was Eric Dillon, a financial advisor who now runs funds that invest in hedge funds and other assets. Nick and Eric had invested about the same amount as I did. Jeff issued each of us five-year warrants to purchase another fifty thousand shares at $1 per share.

As 1996 began, I had become a venture capitalist, and Jeff's dream of an online bookstore had become a reality. Seattle, the bookselling and publishing industries—and for that matter, the world—would never be the same.

1

Opportunities and Challenges of Cities

Seattle in 1995, even with the twenty-year presence of Microsoft, was not at the top of anyone's list of cities destined to become one of the top two technology hubs in the world in the next twenty-five years. How did this come about? How can other cities benefit from Seattle's model?

Some have dismissed Seattle's success as an accident of birth, since Bill Gates and Craig McCaw, the wireless innovator and founder of McCaw Cellular, were both born in Seattle. But Gates and his partner, Paul Allen, might not have returned in 1979 from Albuquerque, where they initially started Microsoft in 1975, had they not seen Seattle's early promise as a place to build a company and attract talented workers. By the 1980s, McCaw clearly benefited from the early elements of Seattle's tech ecosystem.

Jeff Bezos and his wife, MacKenzie, could have picked any city in the country to start Amazon, including staying in New York or heading for the Bay Area, which they considered. But they chose Seattle not only because, as Bezos has said, they believed that they could find the necessary tech talent to hire in Seattle, partly fed by Microsoft, but they could also benefit from the growing strength of the University of Washington's Computer Science and Engineering Department.

They additionally recognized Seattle's openness to outsiders, whether immigrants from New York or Bangladesh. It is often argued that Bezos chose Seattle because Washington has a smaller population than California that would be subject to a local sales tax, but Bezos has never confirmed this. If taxes had been the driver, he should have chosen nearby Portland, Oregon, which has a smaller population than both and no sales tax.

From my office on the thirty-fourth floor of the DocuSign Tower in downtown Seattle, I look out over Puget Sound, its surface dotted with boats and ferries carrying people to and from Bremerton and Bainbridge Island. Along the city's touristy waterfront, I see the piers where in 1903 my grandfather boarded a ship for the Alaska gold rush. One hundred years later, it would be a different gold rush—the Internet gold rush as it became known in the dot-com days of the new millennium—that would reinvigorate Seattle.

What attracted that Swedish immigrant, and so many others, to the mountains and Salish Sea of Seattle? At the turn of the nineteenth century it was often location and opportunity. The young city was bordered east and west by mountains covered with timber growing to the saltwater's edge. Its seas teemed with fish. And, not least, the region was both remote and undeveloped—characteristics attractive to people who believed they could succeed by forging their own paths. With the discovery of gold in Alaska, Seattle became the perfect departure point for thousands of aspiring millionaires—including my grandfather. As they kept coming, Seattle prospered. On the West Coast, only San Francisco rivaled the city as a busier port of trade.

Most prospectors never found gold, but it didn't take long for them to realize the region's other potential. Its abundant timber could answer a national demand for housing, and its vast schools of fish could feed an ever-more populous world. Forty years later, as the firs and the fish declined, Seattle became Boeing-town, churning out World War II aircraft and ships, then branching into commercial jets and aerospace.

One hundred and twenty years after my grandfather's arrival, I see a Seattle that has transformed yet again. The city known for its Skid Road and gold rush saloons, Boeing and World's Fair Space Needle, has evolved into one of the biggest, busiest, high-tech hubs on the planet. Of the five most valuable companies in the world—Apple, Microsoft, Amazon, Google, and Facebook—two are headquartered in Seattle. None are headquartered in New York City, Chicago, Los Angeles, or anyplace aside from the Bay Area. And the Bay Area itself is now a major exporter of people to our midsized city.

Seattle, like the Bay Area, is a leading center of innovation. More than two hundred thousand tech workers now call Seattle home, contributing to one of the fastest-growing populations of any major U.S. city. Nearly 60 percent of residents have bachelor degrees, or higher (the highest rate for a major city), and median salaries hover around $102,000 (more than $120,000 if you count just tech workers). No wonder that thousands of millennials have flooded into Seattle in recent years.

Seattle has become the second-most important technology city on the basis of its breadth and depth of technology expertise—with Amazon Web Services, cloud computing, e-commerce, wireless connectivity, artificial intelligence, commercialization of space, quantum computing, immuno-therapy, global health, and Microsoft's software operating systems and pro-ductivity tools.[1] Seattle is the clear number one in many of these categories. If you want to be a leader in these fields, Seattle is a good place to move.

The tech scene has been complimented by the rise of Seattle's global brands like Starbucks and Costco. The Port of Seattle and Tacoma is one of the largest ports in the world. But clearly tech drives our modern economy, as signified by the recent renaming of my forty-seven-story building from the Wells Fargo Tower to the DocuSign Tower.

∾

Amazon has about sixty thousand employees on its urban campus in downtown Seattle. Microsoft has sixty thousand employees on its more traditional suburban campus on Seattle's eastside within Redmond's city limits. Amazon's Seattle campus includes over thirty buildings intermin-gling other tech and biotech companies, research institutions, restaurants, and retailers. Institutions in the area focused on healthcare include head-quarters for the Gates Foundation, the Fred Hutchison Cancer Research Center (which has created numerous biotech spinoffs) and PATH (a global healthcare institute). The University of Washington (including the Allen School of Computer Science and Engineering and UW Medicine with its hospital and research labs) sits just a mile away as does the Allen Institute of Artificial Intelligence.

Add to them the hundreds of new startups launched in Seattle every year, our twenty or so local unicorns (tech companies headquartered here worth over $1 billion), and more than 140 global companies that have opened significant software engineering offices here, employing tens of thousands more developers, programmers, and engineers.

Google's six thousand employees in the Seattle area are split between Seattle and its eastside suburbs as are Facebook's five thousand employees.

T-Mobile, the third largest cellular company in the United States was founded here and is headquartered on the eastside with twelve thousand employees. Salesforce, when it bought Tableau in 2019 with over two thousand employees, announced that Seattle was its HQ2. Other global tech companies such as SpaceX, Apple, Alibaba, Tencent, and Oracle are growing their local offices. Seattle is positioned to host lots of HQ2s.

Amazon also is developing an urban campus in nearby Bellevue for twenty-five thousand additional employees on Seattle's eastside. Ironically, Amazon's HQ2 has turned out to be not New York City/Queens but Seattle's eastside.[2] A light rail system is under construction connecting Amazon Seattle, Amazon Bellevue, and Microsoft Redmond.

Downtown Seattle, however, also shares major problems with downtowns of other big tech cities like San Francisco with homelessness, high home prices, fraying public safety, traffic, and anti-tech city councils which are causing some tech companies and workers to move out of big cities. These forces are driving some tech workers to dream places like Sun Valley or Montana but mostly, if they move, it is to Seattle's eastside, somewhat comparable to Silicon Valley's south bay area—but Seattle's suburbs are less crowded, less expensive, and more livable.

If our rise cannot be attributed to luck or to our rainy weather, the question remains: How did Seattle grow into such a prosperous, job-filled tech center?

My answer is based on the concept of an economic flywheel, which I discuss in detail in chapter 2. But briefly, the flywheel serves as a model or metaphor for generating momentum that creates urban economic success. The economic flywheel begins with a city attracting talented people whose entrepreneurship, risk-taking, and inventions create businesses that provide jobs and wealth for the community. This attracts more talent, starting more new businesses, and so the flywheel spins faster and faster. In this modern era, the flywheel relies on the tech revolution that provides power for economic growth. Invention and entrepreneurship accelerate the flywheel. In the following chapters, I discuss how Seattle built its economic flywheel, suggest how other cities can do likewise, and point to some inspiring examples of cities you might not expect that have already begun to reinvent themselves.

In Seattle's flywheel, the initial startups were a few small IT and biotech companies in the 1960s and 1970s, including Microsoft. Most faded or were acquired, but even those that failed helped to build an initial tech ecosystem with tech workers and expertise to fuel the beginnings of an economic flywheel. Much the same process was going on in Silicon Valley and Route 128 outside Boston, but the valley soon eclipsed Boston and Seattle.[3] Boston and the Bay Area both benefited from great universities with strong tech and biotech departments, but it was not until the 1990s that the computer science department of Seattle's one major university, the University of Washington, under the leadership of Ed Lazowska, began to catch up.

The booming tech economies of Seattle and the Bay Area, however, should also serve as something of a cautionary tale—for the powerful economic flywheels in both places have not been sufficient to overcome the civic problems that plague major cities. Indeed, economic success has to some extent made these longtime social challenges worse. For all their wealth and brainpower, Seattle and the Bay Area have not been able to solve the riddles of traffic congestion, floundering public schools, increasing numbers of homeless people, and mounting worries about public safety. For this reason, I posit that successful cities need a second flywheel— a livability flywheel.

In its simplest form, a livability flywheel consists of talented residents who provide leadership for successful local governments and civic institutions. At the same time, talented entrepreneurs, inventors, and executives of new and growing businesses create jobs and wealth for the community. Together, governments and businesses are responsible for creating and preserving livable downtowns and solving civic problems. In turn this attracts more talent that accelerates the flywheel.

In part 2 of this book, I explore the successes and challenges of businesses and cities working together and the need for new thinking about building and integrating livability flywheels.

There is overlap between the two flywheels with several common elements. Both depend on a dynamic, growing business community that produces jobs and wealth. Both depend on talent that produces innovation and leadership for government and businesses. And both depend on a livable city that attracts talent and businesses.

Of course, these flywheels can break down, and this is what has happened to the livability flywheels in big cities. For whatever reasons, the challenges of these social issues are overwhelming city governments, and these governments and businesses are not working together to solve the problems. Ultimately, the breakdown of the livability flywheel endangers the economic flywheel.

The relationship among technologies, economic progress, and the ways both could be harnessed to help solve our urban problems is the thread that weaves throughout this book. Our modern economy and these flywheels depend on the major tech innovations of the last fifty years. The development of computers and software were the most important innovations during the last half of the twentieth century, as we went from mainframes to minicomputers to portables, laptops, tablets, and mobile phones. Bill Gates and Steve Jobs led the revolution for making it possible that personal computers could be put on every desktop in offices and homes.

The year Amazon was launched, 1995, was a watershed year for two of the most important innovations of our era—wireless connectivity and the internet. AT&T acquired McCaw and LIN Broadcasting with their nationwide mobile phone network, and Netscape's browser and IPO accelerated the use of the internet by businesses and consumers. Gates's email of that year to all his employees urging them to integrate the internet into all their products and services further signaled the internet's coming of age as Microsoft jumped on board. Also in 1995, Microsoft launched Windows 95, enhancing the use of computers.

Netscape, following its very successful IPO, faded from view as its CEO Jim Barksdale gave up the booming consumer market in the face of challenges from Microsoft and shifted his strategy to business enterprises. Fortunately for the Netscape management and shareholders, AOL bought Netscape, and then pulled off one of the most favorable sales of all time to Time Warner in 2000 for $165 billion, thereby enriching the AOL stockholders as well as former Netscape stockholders who had retained their AOL stock.

The disruptive power of the internet was highlighted by the fact that Google was launched by two Stanford graduate students and Facebook by a Harvard undergraduate from his dorm room. Google became the world's most powerful source of information, and Facebook changed the way people share their lives with others.

In the mid-1990s, analog cell phones were replaced by digital devices, so that people could communicate via text message and begin to access information from the internet. Most of us were carrying a BlackBerry, but we switched to Apple and Google phones when BlackBerry failed to continue to innovate and take seriously the launch of the iPhone, which combined voice, internet search, video, and music in an integrated device. The iPhone was the teleputer, turning the way we talk to each other into a powerful mobile computer for emailing, texting, searching, listening to music, watching video—just about everything we want to do. We feel naked if we leave home or office without our phone. The next evolution of communication devices is the integration with other interfaces for communicating and accessing the internet via Alexa voice in the home, office, and auto; Apple's watch and ear buds; VR/AR/MR; touch; and coming back—smart glasses.

Corporate computing, although advancing in speed and complexity with innovations in software and storage, was generally conducted by machines located on the premises of companies until Amazon launched the forerunner of its Amazon Web Services in 2006, providing computer services on demand by the second and fraction of a second connected by high-speed fiber.

The importance of the impact of these new technologies on our lives can be measured in many ways. An average of 3.5 billion searches are conducted on Google each day, with many of us searching dozens of times a day. Approximately 1.8 billion people log on to Facebook daily. Estimates are that Americans send and receive an average of over fifty text messages on their phones daily, plus sending and receiving dozens of emails. Over three hundred billion emails flood the world daily. Purchasing on Amazon has become the way of life for hundreds of millions, and millions sell goods and services on Amazon.

The increasing importance of technology can be vividly seen in the change in the identities of the leading companies in 1995 compared to today. In 1995, the five most valuable companies were GE, IBM, Coca-Cola, Exxon, and Merck. Today Exxon is number ten, and none of the others are in the top twenty. Think of how weak our economy would be if we were relying on the leading companies of 1995 for our future. In 1995, Apple wasn't in the top fifty; Amazon was launched in 1995; Google didn't exist until 1999; and Facebook was founded in 2000. Will today's top five still be on top in 2045? Unlikely, based on history, but these five companies have strong cultures of innovation and risk-taking that, if allowed to continue in cloud computing, artificial intelligence, and healthcare, may keep at least several of them on top.

Beyond software, the internet, and wireless communications that continue to drive our economy, my top nominations for the life-changing innovations of at least the first half of the twenty-first century are artificial intelligence and biotech. The applications will change human history. And the second half of the twenty-first century? Fearlessly, I suggest quantum computing and outer space.

I must concede that technology forecasters have a notoriously poor record. IBM CEO Thomas Watson is said to have predicted in 1943 that

"There is a world market for maybe five computers."[4] Some thirty years later, Ken Olsen, founder of Digital Equipment Corporation, opined in 1977, "There is no reason anyone would want a computer in their home."[5] And in another famously bad prediction, AT&T in the 1980s commissioned a study of the potential cellular market that concluded there would only be nine hundred thousand phones sold annually worldwide by 2000.[6]

But as a venture capitalist, part of my job is to identify new technologies that could be important to people and businesses and then to invest in new applications of those technologies. In doing so, I have benefited over the years from several technologists who were unusually farsighted in grasping the power of new technologies and placing big bets on them.

Among my guides is Craig McCaw, who in the early 1980s, as a young cable television entrepreneur, envisioned a big future for mobile phones. He foresaw that bulky, sometimes-connecting car phones would be replaced by small powerful mobile phones providing connectivity anywhere, anytime. Craig commissioned no marketing study but acted on his personal insight that everyone would want to be able to connect at all times. He sold the family cable company to invest in wireless licenses beginning in the mid-1980s.

Craig's wireless vision was shared by futurist George Gilder, my close friend and former housemate at Harvard College, who coined the word *teleputer* in 1989 as a device "as portable as your watch and as personal as your wallet. It would recognize speech and navigate streets, collect your news and your mail."[7]

I was fortunate to be hired by Craig in 1990, helping him and his team build out the first national cellular wireless network and experimenting with free long distance. I also led an effort to introduce early wireless data services, which initially was limited to stock prices and sports scores.

At about the same time in the early 1980s as Craig was foreseeing the future of wireless, Bill Gates had the audacity to foresee a computer on top of every desk and in every home. I never worked for Microsoft, but I have worked closely with many of their leaders during my career. Microsoft's leadership in software infuses the air in Seattle.

And, of course, I invested in Jeff Bezos's Amazon and served on its board for twenty-three years, beginning with Bezos and John Doerr in 1996. My understanding of the power of inventions and the future of technology has greatly benefited from being part of the inner workings of Amazon, participating in Jeff's vision that we are still in "day one" of the internet. Amazon's commitment to innovation over all these years is unrivaled.

I also benefit from a friendship with Craig Mundie, currently a senior advisor to the CEO at Microsoft and formerly senior vice president, chief research and strategy officer, and head of Microsoft research. Mundie is a leading thinker on artificial intelligence, quantum computing, medicine, and proteomics.

I also have been inspired and educated by the inventors and entrepreneurs who come through Madrona's doors each week proposing to start tech companies. And I have learned from my fellow managing directors, venture partners, and younger investment professionals at Madrona.

There are those who would dispute technology's forward-looking scenario and its promised benefits. They argue that none of our recent advances will be as transformative as electric lighting, indoor plumbing, cars, and airplanes.[8] But I find it difficult to argue that machine learning/deep learning/AI is not going to change our lives in fundamental ways. Likewise with the biotech revolution. What is important to me is whether the application of a newly invented or refinement of an existing technology substantially changes people's lives. The development of immunotherapy to cure cancer is a fundamental change, even if it is an extension from Mendel's peapod experiments in the 1850s. Autonomous vehicles are a fundamental breakthrough that will significantly change our lives by improving the environment, freeing us from tedious driving, and saving tens of thousands of lives every year. Together, these technologies and the tech companies they beget will transform our daily lives and invigorate our cities and towns.

The U.S. economy and its people have been huge beneficiaries of the new technologies and the creation and growth of tech businesses over the past fifty years. Before COVID-19 struck, the United States was experiencing the lowest unemployment since 1969.[9] Beyond short-term impacts of recessions, financial crises, and even COVID-19, the tech revolution is raising our standards of living and offers once-inconceivable possibilities for solving existential challenges like global warming and life-threatening diseases.

George Gilder and others measure the improvements in our standards of living in terms of "time-price"—"the price of any good or service in the hours and minutes it takes an average worker to earn the money to buy it." Thus, "while population increased 71% between 1980 and 2019, the time-prices of the key commodities supporting human life and prosperity dropped 72%."[10]

The growth of tech industries coincided with the collapse of U.S. manufacturing, as its jobs moved to Asia attracted by lower wages and foreign entrepreneurship. As a consequence, the creation of much of America's wealth shifted from the manufacturing centers in the Midwest like Detroit to cities on the coasts such as Boston, the Bay Area, and Seattle, as documented by Enrico Moretti, a professor of economics, in his groundbreaking book *The New Geography of Jobs*.[11]

In 1950, Detroit was the third wealthiest city in America, with thousands of well-paying middle-class jobs that didn't require a college degree. The jobs of the tech industries, however, required well-educated populations that were developing in coastal cities. Beginning in the late 1950s, Detroit's population began a steady decline that continued into this century. From 2000 to 2010, Detroit's population declined by 25 percent, Cleveland 17 percent, Cincinnati 10 percent, and Pittsburgh, Toledo, and St. Louis each 8 percent.[12] This further decimated their downtowns, which had already suffered from people and businesses moving to the suburbs.

Richard Florida, who coined the phrase "creative class" and was a vocal cheerleader for the economic benefits that would come through attracting diverse thinkers to our coastal cities, now tempers his early enthusiasm.[13] We have overdone it, he says, creating a few "superstar cities" with yawning class divides—and brain drain everywhere else. "As our largest urban centers become increasingly expensive, unaffordable, and divided, they price out and drive away the very diversity that powered their innovativeness and growth to begin with," he writes, labeling this fracture the "central crisis of our time."[14] Our problem, however, is not in attracting too many creative people to tech hotspots—it is in failing to find ways to apply their creativity and wealth to solving our social problems. At the same time, we both agree that there are healthy trends of tech talent spreading to many other cities.

The new tech economy was created in the United States because our free enterprise system allowed individuals to innovate with new technologies and business models and to earn rewards for their efforts that could be recycled into new businesses. Many early tech businesses also were able to leverage the federally funded research resulting from the Cold War and benefited from loose governmental regulation.[15] Thousands of entrepreneurs, inventors, and business leaders contribute to creating and advancing new technologies, and the world's consumers and businesses find benefits in the new tech products and services. The new tech companies also pay well and have an unusually high multiplier effect of five jobs for each tech job,[16] supporting a wide range of nontech jobs in their communities.

Heading into the third decade of the twenty-first century, new forces are spurring iterations of technology that will certainly shape the U.S. economy. Of necessity, COVID-19 has forced us to learn how to keep in touch with our loved ones and business lives through Zoom and other video telecommunications services. Following our conquest of COVID-19, I believe that it is important that we not entirely return to our old ways but draw ideas and actions from these forced disruptions to reimagine our future.

Zoom and Teams, even though lacking the benefits of physical proximity, have their own virtues in being able to bring together a broader range of employees conveniently and quickly from wherever they are located. Some companies have found that weekly Zoom meetings among core executives and even board members are beneficial and that it is easier to limit them to a defined time period such as one hour—which to anyone who has participated in a long drawn-out physical meeting is of no small importance. On the other hand, Satya Nadella, CEO of Microsoft, has noted the benefits of physical presence: "What I miss is when you walk into a physical meeting, you are talking to the person that is next to you, you're able to connect with them for the two minutes before and after."[17] He has also expressed concern for the social isolation that occurs when an employee works exclusively from home.

It will take some years before we fully know the extent of the trend toward working remotely, but with tech leaders such as Mark Zuckerberg, CEO of Facebook, speculating that fifty percent of their employees might work remotely, the impact will be significant.[18]

An increasing number of companies are following GitLabs's example—this eight-hundred-person company was 100 percent remote even before the pandemic. When I talked with one of their executives, he cited the benefits of having a wider geographic base from which to hire talent and the reduced overhead expenses. Part of the money they save from real estate they invest back into their employee travel budget so that everyone can regularly travel for face-to-face meetings with fellow employees and clients. GitLab also has well-developed techniques so that employees can informally network with one another online.

Nine months into the pandemic, I spoke with the leader of a major law office and another at a large accounting firm. Both said the experience of operating remotely has led them to reassess their need for office space. Suddenly, it is clear that every employee does not require a private downtown office that stands empty if they worked from home or at a suburban satellite office. Another business stressed to me that they have provided all their staff who work part- or full-time from home with high-quality computers, video screens, and broadband access. Our workplace norms have been recalibrated and likely will be forever altered.

It appears likely that we will see lots of experimentation among companies over the next several years in varied combinations of remote and in-office working. Most companies will recognize the importance of being physically together for inventing, mentoring, and socializing. These do not require full time attendance in an office, however, so many companies will be allowing more flexible work schedules but requiring that most workers spend three or four days a week in an office.

Some workers will have "co-primary homes" where they split their workdays, sometimes near their downtown office and sometimes near a satellite office or remotely.[19] Similarly, there may be co-primary offices.

Google has said that employees should live within commuting distance of their assigned office and expect to be in the office three days a week. Salesforce has announced three ways employees will work: flex—most workers will work one to three days a week in the office for team collaboration, customer meetings, and presentations; fully remote—employees who don't live near an office or have roles that don't require an office will work remotely full-time; and office based—the smallest population will work from an office location four to five days per week if they're in roles that require it.[20] Amazon has announced that its plan is to "return to an office-centric culture as a baseline . . . [because they] believe it enables us to invent, collaborate, and learn together most effectively."[21]

The definition of "assigned office" is expanding for many companies as they open satellite offices, sometimes in nearby cities and suburbs but increasingly across the United States and also in Canada, Europe, and India. Seattle, as I have mentioned, is a prime beneficiary of companies opening satellite offices with over 140 located in greater Seattle. Other cities are also increasingly the beneficiaries of tech companies opening satellite offices. Amazon, Google, Microsoft, Apple, and others are opening tech hubs in major tech cities like Boston, New York, and Austin and also in cities such as Durham, Denver, Phoenix, and Portland. Austin is becoming an important tech center with major offices and facilities for Apple, Dell, Facebook, and Oracle. In fact, Oracle announced that it is moving its corporate headquarters from the Bay Area to Austin.

In addition to cities benefiting from new satellite offices, they are experiencing a net inflow of tech workers working remotely for major tech companies. As you will read later, Tulsa has had a program for several years that pays about four hundred tech workers $10,000 to move to Tulsa.[22] Now workers are moving to many cities and towns without needing payment and some are accepting lower pay from some major tech companies in recognition of lower living costs. A LinkedIn study showed that the Bay Area lost net workers in 2020 to cities like Seattle, Austin, Portland, Sacramento, and Denver (in that order).[23] At the same time, the Bay Aarea gained a net of workers from other cities, like New York, Los Angeles, Chicago, and Boston, as well as immigration from India. It's impossible to predict the full impact on our cities, but I have little doubt that remote work, education, and medicine will become the norm for an increasing number of people. Following our conquest of COVID-19, I believe that it is important that we not entirely return to our old ways but draw ideas and actions from these forced disruptions to reimagine our future.

Cities and towns should take advantage of this influx of satellite offices and tech workers by investing in their flywheels, with support for angel investors, incubators, entrepreneurs, and startups. These new tech workers can accelerate local flywheels and help build tech-based economies.

COVID-19 also prompted greater cooperation between private industry and government, something we must build on going forward, as I outline in later chapters. The pandemic's acceleration of drug design and approval shows it is possible to eliminate a good deal of the red tape and bureaucratic delay that too often stymie forward motion. The increased use of contact tracing and other technologies that infringe on personal privacy, even though controversial, suggest that it's possible to balance society's needs and personal privacy.

In this book, I advocate policies in which businesses and cities support each other in building robust economies that provide employment opportunities for everyone, and each is supportive of the others' use of technologies and old-fashioned civic engagement to solve major social problems, including a greatly improved education for all, an end to homelessness, enhanced public safety, a twenty-first century transportation system, and increased availability of affordable housing. This book can't solve all these problems, but I intend to point out ways that technology will be able to help and advocate for businesses, nonprofits, governments, and individuals to each make contributions for our benefit.

2

Foundations of the Economic Flywheel

Among dozens of cities that aspired to be one of the top tech hubs, how did Seattle become one of the two leading centers of technological invention in the world? Why did Jeff Bezos decide to base his internet bookstore here, rather than in Silicon Valley or staying in New York City? Why did home-grown entrepreneurs like Craig McCaw and Bill Gates set their roots down here? Even before them, the companies that gave us the first commercially successful ultrasound imagers, heart defibrillators, and barcode readers sprang up in Seattle.

Why not Minneapolis, Memphis, or Minsk? Why does Seattle continue to be a place that nurtures the development of breakthrough technologies? I've spent a long time thinking about this, working with tech companies of all sizes and quizzing colleagues.

All roads lead back to our economic flywheel.

Strictly speaking, a flywheel is a mechanical device, a wheel to which force is applied, creating a rotation that releases energy. Without additional force to turn it, friction will gradually slow the wheel to a halt. But if sufficient new force is applied, the wheel spins faster and faster, creating still more energy that keeps it spinning. Flywheels have been around for at least ten thousand years—potter's wheels are flywheels. James Watt used a flywheel in his eighteenth-century steam engine that powered the Industrial Revolution.

Writers, too, have found inspiration in the flywheel. *Poetry* magazine uses its mechanism as a metaphor for the way "energy, movement and potential" generate new ideas. And in her book *The Writing Life*, Pulitzer Prize–winning author Annie Dillard advises writers everywhere: "Get to work. Your work is to keep cranking the flywheel that turns the gears that spin the belt in the engine of belief that keeps you and your desk in midair."[1]

I was first introduced to the notion of a flywheel as an engine for growing a business at a meeting of Amazon's board of directors in the fall of 2001. We were in the midst of the dot-com bust, and the company was losing lots of money. Jeff Bezos had assembled his S-team of ten senior executives and board of directors to hear from the management consultant Jim Collins, author of the bestselling book *Good to Great* (2001). Collins talked about the concept of a business-centric flywheel, which he explained as the effect of numerous small initiatives acting upon each other. Like compound interest, he said, the growth would be slow at first but gain momentum over time.

The economy was highly stressed. The NASDAQ had peaked in March 2000, and now, with the dot-com bubble burst, a recession had set in. Financial markets that had been wildly positive about internet companies during the late 1990s, showering the sector with venture and public capital, had grown jittery—especially bondholders, and Amazon had raised several billion dollars through bond issues. The company's stock fell from over $100 a share to $6.[2] In May 1999, *Barron's* headlined a front-page story "Amazon.Bomb," and two years later *Washington Post* columnist David Streitfeld quoted a financial analyst who said he "expects the Internet retailer to run out of money . . . later this year."[3]

But Bezos was focused on the long term. He'd invited Collins to talk to the board and senior managers about sustainable growth, so Collins drew for us an Amazon-specific flywheel (figure 2.1). It showed a virtuous cycle, where low prices brought in customers, and more customers allowed Amazon to add more products for sale, the volume reducing prices further and attracting still more customers in a gradually accelerating flywheel. Bezos liked this concept, and the flywheel—with a few Bezos modifications—became a key part of his guiding philosophy at Amazon. He often uses it in employee meetings and public presentations.

The Amazon flywheel is centered on growth. To generate momentum, it begins by providing buyers a marketplace with great product selection, convenience, and low prices. As consumers are attracted, traffic increases

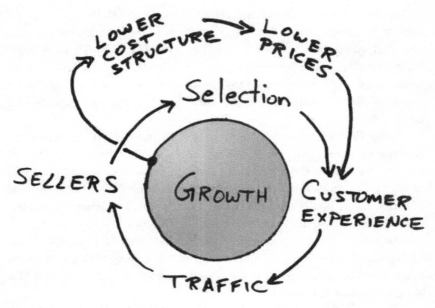

Figure 2.1
Amazon flywheel *Source*: Amazon.

and Amazon can offer more products at lower prices, improving the customer experience, which attracts still more customers. As Amazon grows, the flywheel spins faster and faster.

Amazon also uses innovative business processes—such as opening the site to third-party sellers. They appreciate the ease of selling and shipping products through Amazon. The company benefits from a vastly increased inventory, and competition among the sellers lowers prices. The flywheel concept quickly became integrated as a guiding principle at Amazon, which then conceived new flywheels for its Prime and video divisions.

As I began to write this book, in an effort to explain the increasing momentum in Seattle, I thought back to that afternoon with Jeff and the Amazon team. Board members in the room, in addition to Jeff and myself, were venture capitalist John Doerr, then Gates Foundation CEO Patti Stonesifer, and Intuit founder Scott Cook.

At my firm, Madrona Venture, we talk about the three pillars of Seattle's innovation economy: invention, entrepreneurs, and funding. As I pondered, it became obvious that the most accurate and dynamic way to think about Seattle's economy is with the concept of a flywheel. In Seattle's

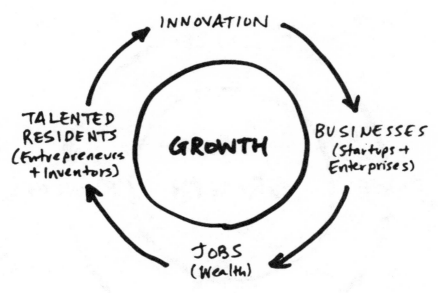

Figure 2.2
Innovation flywheel *Source*: provided by the author.

version, it begins with talented people—entrepreneurs, inventors, and leaders—who create innovations which begets startups. Startups spawn successful companies that then attract talented people who create innovations. Successful companies also generate wealth and jobs, which provide the energy to support the new startups that those creative people inevitably launch. And so the cycle continues. As more talented people move to Seattle, more innovations give rise to new companies, which attract more creative people, and the flywheel accelerates (figure 2.2).

This is a useful way to look at the growth of Seattle's high-tech economy and a model for other cities to emulate when growing their own economies. But how did this flywheel get going *before* we had lots of creative people, inventions, capital, startups, and successful companies? How did Seattle become a world leader in aviation, software, wireless communication, biotechnology, the cloud, and online retail? How did we become a leading place for innovation of all kinds?

Pioneer + Resource Driven

Figure 2.3
Resource flywheel *Source*: provided by the author.

The way I see it, Seattle has had two economic flywheels—one that generated its early growth and one that generates momentum today. What the two have in common is a beautiful setting with plenty of opportunities that attract pioneering, entrepreneurial types and keep them here.

Seattle's earliest ingredients for a flywheel are its location and natural resources (figure 2.3). You could call that the power of place. In the mid-nineteenth century, it was a remote place, with few residents but plentiful timber and fish. The geographic location of this muddy new town—on the shores of the Pacific, ringed with snow-capped mountains—was itself a magnet for people with dreams. We now refer to it as the Mount Rainier effect, a shorthand for the heart-stopping vistas we use as a recruiting tool even today. But the real draw for Seattle, originally and still, is opportunity.

Early Seattle was built largely from the money and jobs generated by the region's abundant salmon, oysters, and trees. Many entrepreneurs came West to make their fortunes, or at least a decent life, from these resources. Others founded companies that serviced those industries. That burgeoning economy provided the capital to build Seattle from a nineteenth-century fishing and logging town into a bustling center of trade, shipping, and commercial fishing by the 1920s. Some of those resource-based startups, like Weyerhaeuser, grew into global brands, and in turn created new wealth, attracting new workers.

To people on the East Coast, Seattle's relative isolation from the rest of the United States, at the far end of nowhere, looked like a disadvantage. But it was actually a strong attraction; the Pacific Northwest was a place you could make yourself over. You could move out and start a whole new life, succeeding or failing on your terms, through your efforts. One of those who tried was my paternal grandfather. He failed to find his fortune in the Alaskan goldfields, but he had the instincts of an entrepreneur, an appetite for risk, and the drive necessary to find his way in the timber industry.

A resource-based flywheel is, however, a dead end. The exploitation of resources does not beget more resources—generally, it depletes them. Timber, fish, and gold could never power the long-term growth of a Pacific Northwest economy. Trade, likewise, was alone not enough to sustain Seattle. Though city officials liked to say Seattle had a natural advantage over California in trading with Japan and China because it was ever-so-slightly closer, that one-day ocean shipping advantage was never great enough to outweigh the Golden State's larger population. And Seattle's advantage as an airline stopover midway between New York and Asia disappeared as soon as planes (many of them built by Boeing) developed the capability to bypass Seattle over the North Pole.

So how did Seattle escape its resource-economy dead-end and move toward a creative-people-and-innovation flywheel? It could have grown incrementally and someday perhaps given birth to a Bill Gates. But its early breakout occurred because of a young man who came to Washington State in 1909 in search of uncut trees. The twenty-eight-year-old head of a logging company based in Michigan, William Boeing, purchased timberland in Grays Harbor County, southwest of Seattle, and made good profits shipping lumber east. That same year, on a visit to the Alaska–Yukon–Pacific Exposition in Seattle, Bill Boeing saw an airplane for the

first time. He took a flight in a small, single-engine plane and was cap-
tivated. More important, he foresaw that aircraft could become the basis
for an enormous industry—even many industries. In 1916, Bill Boeing
founded an aircraft manufacturing company, and its first offering, a sea-
plane, was built partially of Washington spruce timber, which was strong
for its weight.

We will come back to the question of great entrepreneurs and their
importance to economic growth. In this case, it is highly unlikely that Seattle would have become a world aerospace capital without Bill Boeing. And
the Boeing Company stamped Seattle as an early tech city.

When the United States entered World War I, Boeing won a government contract to build fifty airplanes. After the war, the company's fortunes
rose and fell, and then rose again. Boeing focused on commercial aircraft,
secured contracts to provide airmail, and developed a successful passenger airline service as well. As it grew, Boeing naturally attracted smart,
ambitious people to Seattle, engineers in particular. It also found a ready
supply of homegrown talent through local schools like the University of
Washington.

By 1930, Boeing was a conglomerate that included not only Boeing
Aircraft manufacturing but also United Aircraft, which built engines,
and United Airlines. This was too much for the federal government,
and in 1934 when Congress began passing legislation to break up the
company, Bill Boeing retired to his farm in the Snoqualmie Valley thirty
miles outside Seattle. All parts of Boeing prospered as separate companies
after the split-up, with United Aircraft in Connecticut; United Airlines in
Chicago; and Boeing Aircraft based in Seattle, with its vast production
facilities.

Then came World War II, and Boeing grew more, producing military
aircraft. Afterward, new CEO Bill Allen decided civil aviation was the future
and risked everything on the creation of the 707 commercial jet. The rest
is evident in popular history. By the late 1960s, Bill Boeing's entrepreneurial venture had become a global leader in aerospace, commercial aviation,
military missile production, and Saturn booster rockets for space travel.
Seattle appeared in 1970 ideally positioned to give the Bay Area—home to
Lockheed, Hewlett Packard, and Intel—some real competition as a talent
and innovation magnet.

But then Boeing went seventeen months without selling a single plane
to any U.S. airline, laid off seventy thousand workers, and plunged Seattle into its longest, deepest economic trough since the Great Depression.

People fled in droves. "Will the last person leaving Seattle turn out the lights?" bitterly quipped a billboard near the airport. With the Boeing bust and the era's trend of flight to the suburbs, Seattle suffered a net loss of sixty thousand residents between 1960 and 1980.[4]

Boeing was Seattle's original big tech company. But contrary to the belief of many, Boeing was not the source of the city's modern tech economy. Boeing engineers populated a few of the early companies, such as Advanced Technology Labs (ATL) and Seattle Silicon, with former Boeing engineer Gordon Kuenster as CEO, but they did not provide the entrepreneurs nor the software engineers that drove the software companies of the 1980s and 1990s. Boeing provided, however, a tech halo for Seattle's view of itself as a tech city.

By the 1970s, Boeing was known as a metal bender and considered very bureaucratic and hierarchical.[5] Although software was becoming important in the design of their planes, it was not core to Boeing's existence. A perhaps apocryphal story was that when an engineer advocated the introduction of more computers, a senior vice president challenged him to search for the address of a business. The VP pulled out a phone book and proved it was faster than a computer. Unfortunately, the calamities of the recent crashes of Boeing 737 MAX have been attributed to software deficiencies and have proven difficult for Boeing to correct.

One of my goals, when I returned to Seattle and began at the Perkins Coie law firm in 1967, was to work with Boeing. Bill Allen, Boeing's CEO since 1945, who'd made a "bet-the-company" decision in 1957 to launch the 707, had been a Perkins partner. By the 1980s, I became Perkins' lead lawyer on the Boeing account and Boeing's principal outside attorney.

The first Boeing project I worked on in the 1970s was its proposed supersonic airliner. Boeing had decided it was a strategic necessity to develop such a plane, but it involved huge financial and technical risks. My job was to figure out how to structure the project so that an operational or financial failure wouldn't destroy the entire company. When Congress suddenly dropped its financial subsidies in 1971 for the project, Boeing decided the risks were too great to proceed further.

On this and later projects, I got to know many top Boeing executives, including T. Wilson, who had succeeded Allen as CEO. The eight senior Boeing executives had their offices in the squat four-story headquarters building on Marginal Way, across from Boeing Field where planes were built and tested. It was a somewhat art deco–style structure without internal ornamentation except for pictures and models of Boeing planes.

Security was tight. Their facilities were rimmed by tall fences topped with barbwire, and a Boeing badge or an escort was required for entry. Boeing's few top executives, however, could bypass security by driving down a short ramp from Marginal Way that led to an underground garage that had less than fifteen spaces. As a young lawyer, I learned about this from my senior partner, so I always parked there and went up the unsecured elevator to the executive suites. No one ever asked me about it.

When the Boeing bust occurred in 1971, the company began an effort to diversify its way back to revenue and profitability. The executive in charge asked me to help them acquire computer service bureaus in major cities that provided batch processing of data for local businesses, such as accounting firms—a very early precursor to today's cloud services, except that the data was delivered by hand on computer disks. Boeing spent a few million dollars on small acquisitions, but they dropped the idea when it became clear that the business would not scale because of the lack of inexpensive bandwidth for remote processing and the development of relatively low-cost minicomputers by Digital Equipment Corporation (DEC) and others that allowed smaller businesses to process data on their premises. Boeing continued to expand its computer infrastructure and to develop software for its uses, such as designing aircraft and managing supply chains, but it never used its expertise to build a substantial third-party computer services business.

I became involved in Boeing's most bizarre attempt at diversification in 1972 when I was asked to help them evaluate a proposal to use their computer savvy on behalf of New York City's off-track betting business. To my relief, the proposal was ultimately rejected. Boeing also created internal word processing and spreadsheet programs that presaged today's ubiquitous productivity software, and they considered competing with Intel in the design and manufacture of silicon chips. Carver Mead, a professor at Caltech, spent a summer in Seattle teaching silicon design to a group of Boeing employees (he will reappear later in my story as a friend and the cofounder of Impinj, one of the major startups I was involved with as an early investor and board member.) When Boeing abandoned its chip project, several of the employees departed to start a new company, Seattle Silicon, that I represented.

Boeing's heart wasn't really in any of these businesses. What it wanted was for the airplane industry to recover, which it eventually did. But they missed out on what others built into big businesses. It wasn't until decades later that traditional airline and auto industries came to understand that they are as much software as manufacturing companies.

One lesson from Boeing's experience might be, "Stick to what you're good at." But Amazon has built its innovation culture to do the opposite, encouraging its employees to leverage their technical and talent assets, risk-taking, and long-term views to create new businesses beyond internet retail, such as cloud services.

Just as Microsoft mostly failed to participate in the early dynamic growth of the internet in the late 1990s, Boeing missed the early chip and software revolution. Hidden by its commercial and government successes, Boeing had been a traditional manufacturing company all along, with a rigid bureaucracy that stifled invention, entrepreneurship, and efforts to launch new ventures. There had been opportunities to become a leader in semiconductors, business software, and computer services, but Boeing never doubled down on them.

Nonetheless, Bill Boeing's original 1900s entrepreneurialism set the stage for Seattle's innovation flywheel to come, and ironically the Boeing bust helped stimulate a small group of Seattleites to begin building the underpinnings for the tech economy of the future. Although an elitist Lesser Seattle movement argued against committing to a growth strategy, business and community leaders recognized that Seattle needed to diversify beyond Boeing and aerospace to grow the economy. Without fully planning for it, they began building a diversified tech economy that did not depend on a single company.

Beginning in the 1970s, a few entrepreneurs launched several biomedical and software companies. In addition, a few small local venture firms and investment banks were launched to finance startups and their growth. Local lawyers worked to build the financial and tech expertise to keep the tech legal work from being captured by law firms in New York City and San Francisco. Seattle was laying the groundwork for their tech flywheel to begin spinning.

As chief outside legal counsel to the Boeing Company in the 1980s, I had watched a global company struggling to maintain its edge in technology. In the early 1990s, as executive vice president at McCaw Cellular Communications, I got an early ground-floor view of the future of mobile communications and the power of innovation.

In 1995, after becoming one of the first investors in Amazon, I had no idea that the tiny startup would become one of the most valuable companies in history, displacing corporate icons like IBM, General Electric, and General Motors. But I did grasp that the future of successful cities would depend on attracting talent and creating growing technology firms.

It was clear something transformative was happening in my laid-back, blue-collar town. As a former tech executive and newly minted venture capitalist, my investing strategy relied on the belief that technological innovation could transform not only business and industry but also cities and society. I was becoming optimistic that a tech economy might spur a renaissance in American urban life—healthier, more prosperous cities, with innovation creating untold possibilities. Along the way, I thought, we'd also find solutions to the stubborn problems of mass transportation, affordable housing, street crime, and public education.

In retrospect, I think I was right in that tech would create economically prosperous cities, such as Seattle and San Francisco, but, as I will discuss in the latter part of this book, it has proven much more difficult for business and public policy leaders to use technology to solve our social problems. In some ways, the problems have been aggravated.

Beyond Seattle and San Francisco, cities like Boston, New York, and Austin are benefiting from their economic flywheels that are spurring the growth of their technology economies while suffering with Seattle the challenges of creating a livable city. My visit to the heartland in early March 2020, literally days before COVID-19 spread throughout the United States, helped open my eyes that initial tech and livability flywheels are beginning to spin in many cities such as Tulsa, Oklahoma City, and Kansas City. Cities that I had heard labeled part of the Rust Belt are rebuilding their downtowns and waterfronts with new and restored buildings, restaurants, entertainment, and housing, reversing the flow of talent to bigger cities by providing opportunities to local graduates and attracting young people looking for places to raise their kids and build their economic futures. Tech incubators are being created, new companies launched, and capital invested. Local political leaders are rising above the partisanship common in Congress and leveraging technology and businesses to improve the lives of their residents with jobs and solutions to urban problems. If a further catalyst was required, COVID-19 and the increasing trends to working remotely will help accelerate their flywheels by increasing the flow of creative people to these cities.

I hope this book will be a contribution on how to build flywheels that promote economic development and jobs in all our cities and strengthen our abilities to solve urban social problems.

3

Seattle's Flywheels Begin Spinning

During my years as a lawyer, Boeing presented lots of interesting legal work for me, and the people were smart and determined. But from an early date, I had more fun and interest in young tech startups who struggled to raise money and launch innovative products.

My first experience with financing such a company occurred in the back of a taxicab, in 1976, many years before I met Jeff Bezos. Donald Baker, a researcher at the University of Washington (UW) School of Medicine, had developed equipment that could show real-time images of tissue, blood vessels, and organs inside the human body using ultrasound. ATL (Advanced Technology Laboratories) licensed the technology from the UW, and initial financing had come mostly from Sam Stroum, a local businessman who owned a chain of auto parts stores. Stroum and the few other local angel investors recruited as the CEO Gordon Kuenster, a mid-level Boeing executive who was frustrated by the bureaucracy and lack of entrepreneurial opportunity at Boeing. He proved to be the exception to the rule that startup founders did not come from Boeing, turning out to be a driven, talented entrepreneur. When he approached Perkins to represent ATL, my senior partners were reluctant to represent this risky venture but surprisingly acceded to my pleadings. ATL had no money, no salable product, and a handful of employees. We struggled to locally raise the equity capital necessary to hire key employees and refine the technology. A New

York City angel, Fred Adler, heard about ATL and came to Seattle to check it out. He had made lots of money by investing in Data General, an early minicomputer manufacturer based in Boston, so he knew something about tech startups. In Seattle, he met ATL's founders and viewed the technology but without committing.

As I sat with him in the taxi taking him back to the airport, he wrote out a personal check for $50,000 to ATL and handed it to me. "Here," he said, "that should help them get moving. Draw up the paperwork and mail it to me. I trust you." Adler and a couple of his New York friends followed up with substantially larger investments, which paid off for them and Seattle's future when ATL was sold to Squibb Corporation in 1974 for $60 million. Adler and his friends received $2.5 million on their total original investment of $178,000.[1] ATL still operates in the Seattle area as a subsidiary of Phillips, employing over one thousand people making advanced ultrasound machines. The sale price of $60 million sounds like a small amount today, but it produced a nice return for local investors who then reinvested into the tech flywheel. ATL and other early tech companies were also the training grounds for future entrepreneurs, investors, and CEOs.

I learned some early investing lessons from Adler's angel investment. Adler acted quickly and decisively without waiting for every *t* to be crossed. In general, though, you should perform due diligence—which includes nailing down legal documentation—before investing any money. The better lessons for me were to bet on a breakthrough product and the people. And to be willing to take a measured risk. Adler's initial $50,000 risk was relatively small and bought him time to secure other investors, perform a more detailed examination of the business plan, and obtain legal protection.

It was also during this period that Bill Gates II befriended me and became an early mentor. He became known as Bill Senior, as the fame of his son grew. Bill Senior was fifteen years my senior and worked for a competing law firm. Notwithstanding the gulf in our ages, experience, and reputations, Bill Senior treated me as an equal in all our relationships. When he was ninety-two, he and his wife would have dinner with my wife and me at our house. He was still a friend when he died in 2020.

Few remember that Bill Senior had been Seattle's leading lawyer representing early technology startups before Bill Gates III launched Microsoft. Among the companies he represented were Physio-Control, the creator and distributor of the first successful external emergency heart defibrillators, and Interface Mechanisms, or InterMech, the developer of the first barcode readers. Physio-Control went public in 1971 and was acquired by Eli Lilly & Co. in 1980 for $145 million, which equated to $170,000 for every $1,000 the original investors had invested. The local wealth it created further fueled our flywheel.

The founders, employees, and investors of early companies, such as ATL, Physio-Control, and InterMech, would provide the guts of many of the tech companies that would be founded in Seattle over the next twenty years. An example is Peter Van Oppen, who after working at ATL went on to be CEO of Advanced Digital Information Corporation, served on numerous tech boards, and is a partner with Trilogy Equity Partners, a leading venture firm in the Seattle area. With a big assist from Microsoft, founded in 1975, these people and organizations set Seattle's flywheel spinning. My role was as a lawyer, business advisor, and "connector."

Seattle is not only one of the premier centers of innovation in software, the internet, and the rise of mobile phone technology, it is a major center for the development of biotechnology. Here is my insider's view of how that came to be.

Immunex, one of the earliest major biotech companies in the nation, was founded in 1981 by two medical researchers, Steve Gillis and Chris Henney, at the Fred Hutchinson Cancer Research Center (the Hutch).[2] Steve Duzan, its CEO, retained me as its outside lawyer. One of the first things I did with Duzan was to negotiate the license deal with the Hutch so that Immunex would have the exclusive right to develop and market the drugs that Gillis and Henney had worked on there. Duzan and I were novices in the new biotech world, as were most people, but we learned quickly how to structure financings and licenses.

At the negotiating session with the Hutch, we quickly agreed on standard royalty rates and terms. Then the Hutch said that they also wanted to own some of the equity of the company. I asked how much, and they responded "fifty thousand shares." I guess that fifty thousand shares sounded like a lot, but it was the percentage ownership of the company that mattered. Depending on the number of shares that were being issued to founders, employees, and investors, fifty thousand shares was either a big or a small percentage. In our case, it wasn't much, and I quickly agreed to their number. More recently, when the Hutch licensed rights to a new drug technology to startup Juno Therapeutics, the Hutch was much more sophisticated and insisted on receiving shares representing a significant percentage of the initial capitalization and additional shares depending on how well the company performed.

~

Duzan and I traveled the world negotiating sublicense agreements with major pharma companies in Tokyo, Frankfurt, and New Jersey. We learned at each stop. Duzan played the bad cop and I the good cop. One time in late December, after several days of intense negotiations with Beringverke in Frankfurt, Duzan rose from the negotiating table and walked out of the room saying, "Merry Christmas." It was left to me to explain that he had just agreed to the final terms. Somehow it all worked, and Immunex was well-financed. Immunex went public in 1983 and entered into a joint venture worth more than $1 billion with American Cyanamid in 1992. After several financial gyrations, Immunex was acquired in 2009 for $16 billion by Amgen.

At Perkins Coie, we also represented Genetic Systems Corporation, an early antibody biotech highflyer founded by Robert Nowinski and financed by his showmanship and the infamous Blech brothers from Brooklyn, New York.[3] Later on, Nowinski and Henney of Immunex would team up with George Rathmann to found ICOS in 1989, the creator of Cialis. Rathmann had been the cofounder of Amgen and left there to be the CEO of ICOS, where Bill Gates was a major investor. ICOS is a great example of the value of recycling local wealth and entrepreneurial talent into the startup community. It went public in 1991 and was acquired by Eli Lilly in 2006 for $2.1 billion. The Hutch's role in these companies shows how the research at local nonprofit institutions can be leveraged into economic development.

One lesson I learned from my experiences was that you need to be prepared to pivot your business if your original plans are not succeeding. Immunex's initial IL-2 drug failed, but its later drug, Enbrel, made billions. Madrona now calls it finding "product market fit." A particular biotech lesson was the importance of raising significant capital early on the promise of the science so that development and human trials can be sustained, even if there are disappointments along the way.

The other important part of this story is what *didn't* happen with biotech in Seattle. None of Seattle's three largest biotech pioneers, although successful, grew into long-term independent pharma companies but were all bought by big pharma corporations: Immunex by American Cyanamid and ultimately Amgen, which later decided to move its operations to California; Genetic Systems by Bristol-Myers Squibb; and ICOS by Eli Lilly.

~

While investments in biotech startups in Seattle continued, by the mid-2000s there were no large successful local biotech companies to point to. Thanks to the research strength of Seattle's Hutch and the University of Washington's Medical School, however, there have been several new medical breakthroughs that more recently allowed for the formation of successful new companies. In the field of immunology, one of the leading new biotech companies was Juno Therapeutics, which went public in 2015 and finished its first day of trading at $2.7 billion. Juno appears again in a later chapter on important new technologies. Juno was bought by Celgene in 2018 for $9 billion. Most of Seattle's successful biotech startups are acquired by bigger drug companies, but that is likely OK as Seattle spawns an increasing number of successful breakthrough biotechs, and the talent and wealth is recycled into new promising startups.

By the late 1980s and early 1990s, new information technology companies were launching in Seattle, including Aldus, Visio,[4] and Real Networks. Each had interesting entrepreneurs and stories, trained future tech entrepreneurs and employees, and distributed money into the community for the flywheel.

I caught up with Paul Brainerd, the founder of Aldus, in the small town of Glenorchy, twenty-six miles north of Queenstown, New Zealand, in January 2020—just before COVID-19 exploded worldwide. After selling Aldus to Adobe in 1994 for $525 million,[5] Brainerd retired from business but not from the world. He and his wife Debbie devoted their lives and dollars to creating a leading outdoor environmental learning center (IslandWood) on 250 acres they acquired on Bainbridge Island, a twenty-minute ferry boat ride from downtown Seattle. It inspires and educates more than twelve thousand children annually.

Now they spend half of each year in Glenorchy where they devote themselves to supporting the local community by maintaining an ultra-green eco-retreat with lodging, an adjacent campground, and general store—a description that does not do it justice.[6]

As my wife Judi and I lingered over dinner with Paul and Debbie, at a lakeside restaurant overlooking the mountains where much of *Lord of the Rings* was filmed, Paul was happy to reflect on how in 1985 he revolutionized desktop publishing.

After being laid off as a vice president of maintenance and customer service for a tech company, Brainerd invested $100,000 to start Aldus which launched graphical user-friendly software, PageMaker, that made it easy to layout print and graphics on a page. This software turned Apple's floundering Macintosh into a highly successful computer. Determination and serendipity, as is often true, playing roles. Brainerd related that he made a hundred presentations to venture funds before one in Silicon Valley agreed to invest. Although he wanted to launch his software on IBM's personal computer, IBM ignored him, and an Apple salesman, without authorization, delivered him four Macintoshes to test.

Other than Microsoft, the most significant Seattle tech company in the 1990's was McCaw Cellular Communications, the first to tie together a national cellular network. It was also the first to provide digital data over cell phones and to offer free long-distance phone service—both radical innovations that, in the early 1990s, were years ahead of their time. Just as Bill Gates foresaw a computer on every desktop, Craig McCaw foresaw a mobile phone in every hand and worked to build a national wireless network to make that possible.

Arguably, software, wireless, and internet technologies were the three most important innovations that enabled the tech revolution, and Seattle's Gates, McCaw, and Bezos can be credited with playing critical roles in the development of all three.

I participated in small meetings with Gates and McCaw back in 1994 when for a time it seemed as if they might use their mobile and software expertise to build a universal email system. Gates and McCaw shared the ability to imagine where technology was headed. Although Craig was fascinated by new technologies, he never wrote a line of code and thus violated Gates's oft-made comment that only technologists could be successful CEOs of technology companies. At one of our meetings, Gates exclaimed to Craig, "I don't understand how you make your money since you don't f***ing own the technology, but you obviously make a lot of it."

Though Gates invested in some of Craig's projects, a full-blown collaboration between the two remains one of the interesting might-have-beens of the tech era. A greater one is: If Craig had not sold McCaw Cellular and LIN in 1995, would McCaw Cellular be one of the big five tech companies today? Possibly yes, but Craig was cut from a different cloth than Gates, Jobs, Bezos, Zuckerberg, and Page/Brin, and he wasn't prepared to commit for the long-term building of a company.

I worked for Craig during McCaw Cellular's peak years—from 1990, when it purchased 50 percent of LIN Broadcasting, to 1995, when we completed the sales of McCaw and LIN to AT&T. As executive vice president of McCaw and president of LIN, I found that up close Craig was brilliant, bold, stubborn, fearless, deeply private, and altogether original—a man, as a 2002 profile in *Fortune* put it, who "honestly doesn't seem to care what other people think" when it comes to his sometimes wacky-sounding ideas. *Fortune*'s writer asked him to describe himself, and Craig said he was "a well-intentioned introvert" with a "fundamental belief that anything is possible."[7] That's a fair, if incomplete, assessment of this frustrating and formidable figure who did much to make Seattle what it is today.

Craig was a child of great privilege, but when he was still a teenager his family experienced a terrible reversal of fortune that likely steeled him for the obstacles he would have to overcome later. His father, John Elroy McCaw, was a high-flying broadcast and real estate entrepreneur who purchased radio stations in many cities, including in 1953 the New York City station WINS for $450,000. He sold the station for more than $20 million seven years later.

Craig was home during his sophomore year at Stanford University in 1969 when he discovered his father's body in bed. Most of Elroy's fortune was tied up in debt (it would take Craig's mother years to clean it up). After Elroy died, his son began running the cable operation from his college dorm. A year later, when the *Seattle Times* made an offer to buy the operation for $720,000, Craig convinced his three brothers to turn it down. He assumed the role of CEO and began to grow the business, purchasing his first cable company for $50,000 while still in college. In 1976, he changed the company's name to McCaw Cellular Communications, moved it to Seattle, and with his brothers aggressively began to build an empire of his own.

Craig bought cable companies everywhere he could—about twelve operations per year, all of them on margin. In 1981, he sold 45 percent of the company to Affiliated Publications (owners of the *Boston Globe*) for $12 million, giving him more cash to leverage. But he was already thinking about the next big thing.

As Craig told the *New York Times* in 1992, he'd had an epiphany one day back in 1979, while driving the streets of Seattle in his Pinto station wagon and struggling to use what was then the most advanced wireless car phone available. "I was dialing 10 times, sometimes 20 times, even 30 times, just to complete a single, scratchy, disconnects-if-you-turn-the wrong-way phone call. It seemed to me that if it could be done right and in a big way," he said, "cellular phones would be worth a fortune."[8]

When the Federal Communications Commission (FCC) announced that it was planning to issue two cellular licenses for each of the country's 733 markets, one to AT&T and one to an independent company selected through hearings, later lotteries, and competitive bidding, Craig put in applications for Seattle, Portland, San Francisco, San Jose, Denver, and Kansas City. He won all or part of all six. That was only the beginning. In 1987, McCaw sold its cable TV properties for $755 million to focus all of its efforts on the cellular phone licenses. It formed local syndicates with rivals, each party agreeing to contribute any cellular licenses they acquired to a joint venture. McCaw also pursued the winners of the FCC's cell license lotteries and offered them cash, paying some as little as $5 per population covered (POP) but shelling out more and more over time in its aggressive effort toward consolidation. At the time, the prices Craig was willing to pay seemed excessively high, but they turned out to be bargains in the long run. In 1986, McCaw sold its cable TV properties for $755 million in order to focus all its efforts on the cellular phone licenses.

Seeking ever more cellular market share, McCaw launched a hostile tender offer for LIN Broadcasting in 1989. In addition to its TV stations, LIN owned or controlled cellular licenses in New York, Dallas, Houston, and Los Angeles. After months of bitter fighting, LIN agreed to sell McCaw 50 percent of its shares, and McCaw agreed that five years later, in 1995, it would either sell its stake in LIN to the public or acquire the company out-right at a price to be set by appraisal. The effective price McCaw paid for its 50 percent stake—$335 per POP—was, at that point, the highest ever. But when the LIN deal was done, McCaw had become the largest independent cellular services company in the nation.

For all this, McCaw Cellular Communications was still relatively unknown in Seattle. Cellular technology had advanced well beyond the car phone that had so frustrated Craig in 1979. We look back and joke about "the brick," Motorola's DynaTAC 8000X from 1983, which cost $3,995, took ten hours to charge, and offered just thirty minutes of talk time. Contractors used them at job sites and quipped that they could double as hammers. No one (except Craig) was predicting that cellular phone service would become the primary means by which people communicate, receive news, and keep up to date with their lives.

It was AT&T that had invented cellular telephony, and in 1980 the company enlisted management consultants McKinsey & Company to estimate the market potential for cellular phones. After crunching the numbers, McKinsey's analysts predicted nine hundred thousand mobile phone users in the United States by 2000, which turned out to be off by a little more than 99 percent. The real figure was 109 million. Today there are over five billion worldwide users. This colossally bad prediction was likely what convinced AT&T to cede its free licenses to the Baby Bells.

But Craig saw what was to come. One evening in the early 1990s, as he and I walked back to our hotel in New York after a long day of meetings, Craig half-jokingly suggested that we hire models to stroll up and down Fifth Avenue pretending to talk on their mobile phones as a way of puncturing people's reluctance to do this in public. We talked about opening a fancy mobile phone store on Fifth Avenue to make mobile phones into status symbols. (Apple grabbed the ball on that one with its iconic Apple Store at Fifty-Ninth and Fifth.)

The first time I ever spoke to Craig was in 1988 when he called to recruit me for a job as general counsel at McCaw. After meeting with him and thinking about it, I told him that I was impressed with what he and his brothers had accomplished, but if I was going to continue working as an attorney, I preferred the independence and varied opportunities afforded me through my law firm, Perkins Coie.

Several months later, Craig called again, this time asking me to represent him personally regarding a contract he had with his brothers that effectively ceded control of the company to him.

In early 1990, he called once more. He'd rethought his previous offer from 1988 and wanted to revisit it. Would I be interested in joining McCaw as an executive vice president to work with him on strategic matters? I would start as general counsel, but one of my first tasks would be to recruit someone else to run the legal department. By then I had developed a much clearer understanding of cellular communications and the growth potential of McCaw under Craig's leadership. I had been wondering, privately, if I hadn't made a mistake in turning him down the first time. And the position he described sounded more appealing this time around.

Craig designated John Stanton, then vice chair of McCaw Cellular, to negotiate the terms of my employment. He prepared a spreadsheet showing the potential value of my stock options over time and gave me the choice of having their price set at the time I joined or progressively, at the then price each year, as the various options vested. Believing in the company's prospects, I elected to have them priced on the date I joined the company. This proved to be a mistake, since shortly after I started the new job McCaw's stock price plummeted from over $20 per share to about $10.[9] Several concerned insiders sold their holdings at the bottom of the market. But I had seen other young public companies, like Immunex, go through similar declines, so I wasn't worried. Over time, McCaw's stock more than recovered, to the regret of those who sold early. Other tech companies have weathered comparable volatility—most notably Amazon between 1999 and 2001, where I still own many of the shares I bought in 1995. You've got to be prepared to hang on.

McCaw was deeply in debt and highly leveraged when I joined. The capital required for its acquisitions was substantial, and Craig had used every bit of creative financing available to raise it. Craig and his team worked with Michael Milken at Drexel Burnham Lambert, then a high-flying investment bank, to issue over $4 billion of junk bonds. They also sold 22 percent of the company to British Telecom for $1.3 billion. I had analyzed all of McCaw's public financial statements, and although its financial condition looked risky, I was undaunted. Not everyone was so sanguine. They had long-term debt which was rated "junk grade" by Standard & Poor's. While I was still making my decision to sign on, Geraldine Fabrikant of the New York Times wrote about McCaw's "sea of red ink" and the "long shadow" that its debt load was casting.[10]

On my first day on the job, Craig called his top executives together to review the company's strategies going forward. He started by going around the table and asking each of us, in turn, to define McCaw's business. Telephones, everyone said. "No," Craig responded. "We are not in the phone business. We are helping people communicate with each other, whether they use voice, video, text, or whatever other medium is invented in the future." His farsighted vision made a lasting impression on me.

Several years later, Craig and I met with Alan Mullaley, vice president of engineering at Boeing, to propose a joint venture for the launch and ownership of the Teledesic satellite phone network. At one point in the conversation, Mullaley, who later was CEO of Ford, declared that Boeing was not in the airplane business but in the business of delivering people, spookily echoing Craig. Likewise, Jeff Bezos, despite initially framing his vision as the "world's largest bookstore," evolved Amazon to serve as true one-stop shopping, a place that could meet his customers' every product need.

Like Mullaley, Bezos, and Gates, Craig was a highly original, big-picture thinker, unbound by dogma or conventional wisdom. When he looked at the status quo, he saw how it could be disrupted. When he encountered challenges, he saw opportunities that others couldn't imagine. Some of Craig's unusual thinking may have a neurological basis; he simply didn't take in information the way most of us do. It is no secret that Craig is dyslexic, but I didn't realize it at first. I finally caught on after we'd had several meetings in which it seemed clear that he hadn't read my memos. That was true, but he compensated for his difficulty with reading by being a very careful listener.

Perhaps that's why the quality of voice communications was so important to him. He pushed his engineers relentlessly to improve the clarity of cellular calls and was frustrated by their slow progress. Once, he showed us a marketing video produced by Steve Jobs in which parents spoke with their children and grandchildren overseas via a video call on a large color screen. Both the picture and the sound were flawless. Only now as we all learn to use video conferencing is this vision reaching fruition, although lack of bandwidth and processing power sometimes still means the movement of one's lips doesn't match the words and voice, and video sometimes becomes garbled. I have no doubt that if Craig had held onto his company longer, we would have better video calls today. In all probability, Craig would have launched something like Amazon's Echo and Echo Show long before Bezos did.

Introverted as he is,[11] Craig can be fearless when it comes to challenging others and their thinking. When Motorola visited us, prepared to make a presentation with an elaborate slide show, Craig said, "Put away the slides and just discuss your proposal," completely unsettling them and, not incidentally, seizing the upper hand. Another time, after a conference room meeting at AT&T's Bell Labs in New Jersey, which had extremely tight security, Craig led us on an unplanned tour around the building, barging into laboratories and asking scientists to explain the projects they were working on. AT&T executives trailed behind, not knowing what to do.

Craig's refusal to look at Motorola's slides wasn't just a power play—he had a real aversion to PowerPoints though they were widely used by others at McCaw. Jeff Bezos doesn't like them either. Instead, Bezos requires his Amazon executives to submit tightly worded, six-page arguments—single-spaced—that are read in "study hall" at the beginning of any meeting where important decisions are being made. Craig might not read his memos, but he similarly insisted on oral discussion and reasoned debate.

At one point, Craig decided we should all take the Myers-Briggs personality test. He scored as the most intuitive thinker among all of us. I was a distant second. Craig's most extreme public prediction, for which he was ridiculed, was that humans would someday communicate through chips implanted in their brains. Twenty-five years later, Elon Musk, not a small thinker himself, has funded Neuralink, which is working on a "direct cortical interface" for communicating thoughts between people.[12]

I accompanied Craig on trips all over the country in his Falcon 50 and other personal jets to meet with bankers and the CEOs of Baby Bells. His style in those meetings was improvisational and unpredictable, and he had no qualms about putting people on the defensive. Once, when we were in New York, he marched the two of us into Bear Stearns' headquarters on Madison Avenue and East Forty-Sixth Street without an appointment and demanded to see Ace Greenberg, its CEO. When we were finally ushered into the nonplussed executive's office, Craig accused him of short-selling McCaw shares. Greenberg denied it, insisting that he "loved McCaw Cellular." Craig's response was a mild, "We look forward to your support." He had made his point.

Just as Craig did not believe in business constraints, he disregarded ordinary personal limits. If it was a sunny day in Seattle, he might decide to fly us to an island in Puget Sound for lunch so we could eat at a waterfront restaurant. Planes, yachts, and money, in no particular order, gave him freedom. A nomad at heart, Craig believed humanity's natural state was closer to that of the tribes of the steppes of Central Asia, rather than one of farms, villages, and cities—even though he made most of his money providing wireless service to cities.

In July 1990, Ted Turner brought the Goodwill Games to Seattle. Craig chartered a sightseeing boat and invited Turner, Bill Gates, TCI's John Malone, and some other corporate leaders in town for the games to cruise Lake Washington. The lakefront was dotted with impressive houses, some owned by Craig and members of Seattle's traditional business elite, like John Nordstrom. But the billionaires' mansions that line the shore today, occupied by Bezos, Gates, and absentee plutocrats from China, were yet to be built. Seattle was still very much unknown to much of the corporate world.

With his characteristic disdain for rigid management structures, Craig quickly loaded me up with additional responsibilities, including oversight of a mobile phone system for airplanes. It relied on eighty cellular towers on the ground and phones installed in passengers' seatbacks. Craig's brother Bruce had extensive aviation contacts, and I appointed one of our young lawyers, Keith Grinstein, to lead our efforts. They quickly signed up airlines, and at its peak 1,700 planes were equipped with our system. We focused on voice service, partly because cellular data was in its infancy (some of us had prematurely dubbed it an "office in the sky"). Even today voice calls are not a killer use on planes. In the 1990s they were expensive, and those who made them felt uncomfortable talking on the phone next to a stranger. We considered building phone booths on the planes! But it was a losing proposition. We should have focused our efforts on simple data services and pager-like text messaging.

After I had been at McCaw for six months, Craig asked me to add president, chief operating officer, and director of LIN to my duties. McCaw, in acquiring its 50 percent interest in LIN, had agreed to maintain three independent directors to represent the public shareholders. I was charged with aggressively growing LIN, but when McCaw became a part of AT&T I was suddenly in conflict with its executives—and with former McCaw executives who had joined AT&T. AT&T wanted to buy LIN for as low an appraised price as possible.

Each of the local markets that we operated in was run by a McCaw or LIN executive who reported to their respective presidents. After I became president of LIN, the managers of the New York City, Los Angeles, Dallas, and Houston markets reported to me, as did the president of LIN TV, which owed NBC affiliates in Louisville, Indianapolis, Nashville, and Dallas.

Reflecting Craig's sometimes casual management style, I met with him to propose that McCaw work jointly with the Kwok brothers in Hong Kong to acquire a cellular license. I had prepared a lengthy memo based on initial conversations with potential partners. As usual, Craig wasn't interested in the memo, but after five minutes of discussion, he authorized me to proceed. I enlisted John Stanton to marshal the tech team, and after several trips to Hong Kong, our joint venture with the Kwok brothers and a mysterious 10 percent participant from Beijing was awarded the license. The Hong Kong operations grew into a valuable asset for McCaw.

A few months after I joined McCaw, Craig told me that he had begun conversations with the CEO of AT&T, Bob Allen, about entering into a strategic partnership or possibly even being acquired. This surprised me. I'd joined McCaw with the expectation that part of my job would be to grow the firm enough to challenge the Bell companies. Craig asked me to colead the negotiations with vice chairman Wayne Perry. Craig rightly thought that Perry and some of his team were too eager to sell; he wanted me to provide some balance, or you could say, restraint.

AT&T, unlike its children the Baby Bells, did not provide local telephone service and at the time of its breakup had elected to give up rights to a nationwide cellular license. Most of its money came through long-distance phone calls; the acquisition of McCaw looked, I'm sure, like a way back to providing local telephone service. But it was not clear to anyone why Craig McCaw wanted to sell.

He was an enigma to all of us. Although serious when necessary, Craig had an irreverent management style. He saw himself as a disciplined rebel, someone willing to take risks but careful to avoid anything that might endanger his company. At banquets, he was prone to lobbing dinner rolls. In the office, he might suddenly launch a squirt-gun attack. Many of us kept our own, loaded and ready behind our desks to fight back. We printed shirts with "Who are those guys?" from Craig's favorite movie, *Butch Cassidy and the Sundance Kid*.

We had a series of endless negotiations with AT&T at remote locations in small towns like St. Cloud, Illinois. (The secrecy held, though it was pretty obvious that something was going on when we descended on these minor airports with fleets of private jets easily traced to AT&T and McCaw.) One or two AT&T vice chairmen attended the meetings, but most of the negotiations were conducted by their lawyers. Every paragraph was a fight over minutia.

Craig did not attend the negotiations, and Jim Barksdale, McCaw's president, was an infrequent participant. Barksdale was trained as an IBM sales executive and had a folksy, charming style, interspersing his talk with Mississippi colloquialisms. "That dog ain't gonna hunt" was his favorite way to dismiss an idea. Negotiations fell to Perry and me together with our able legal team. In the evenings, I would talk on the phone with Craig from my hotel room to describe what was happening and get his input.

After some months of negotiations of a licensing-type structure, it became clear to us that this was unworkable. Craig asked me to present our case for AT&T buying McCaw at a meeting we had with AT&T's CEO, Bob Allen. After more negotiations, we reached a preliminary agreement that AT&T would buy all of the shares of McCaw for shares of AT&T, but the final exchange ratio was left to the end of the negotiations.

There were two classes of McCaw stock, class A controlling shares held by Craig, his brothers, and a few long-time executives, and class B, owned by the public and most of the employees. AT&T assumed that Craig would favor a deal structure where the controlling shares received a higher price than the noncontrolling shares. But privately, I explained to Craig that although paying a premium was likely legal, it would be viewed as unfair by many public shareholders and, importantly, by McCaw employees who owned only noncontrolling shares. I also told him that I thought AT&T would use it as a wedge in our negotiations to pay a lower total price for McCaw by giving class A's a premium and then insisting on an artificially lower price for the noncontrolling shares—figuring Craig and some of the other executives would not be motivated to resist this. Craig considered this for a couple of minutes and said, "Let's treat all the shares the same." Everyone was surprised at his position. AT&T realized it had lost a negotiating lever, and McCaw executives with control shares realized they would not receive any premium over other employees and shareholders. In the final negotiations, we were able to focus on maximizing a single price with undivided loyalty on our team.

One time Craig said to me that we should ask for some side benefits from AT&T in the negotiations. After selling his cable companies to Jay Cooke Kent who owned the Washington Redskins, he'd always regretted not asking for free lifetime tickets to their football games. "Kent would easily have given them to me," he said. I had a friend who had received free long-distance service for life after he'd retired as an AT&T executive. So Craig and I decided to ask for free cellular service for all top executives and board members at McCaw for life. At the next negotiating session, I casually mentioned this provision. "No problem," the AT&T executives said. So, as instructed by Craig, I added, "globally for all the phones" each of us had. No one asked, but Craig had over twenty phones and I had six. Some years later, a subsequent generation of AT&T executives asked, "Who agreed to this?" and tried to terminate it. But twenty-five years later, we still enjoy free cellular service.

By early November 1993, we were finally close to an agreement, and the McCaw team flew to New York on a Thursday for final negotiations. We told our board that they should be prepared to arrive in the city for a meeting on Sunday to approve the deal. We were encamped in a four-bedroom suite that McCaw had leased for several years, covering the top floor of the elegant Lowell Hotel on East Sixty-Second Street, off Madison Avenue.

On Friday, Craig called a meeting of his top executives. He asked each to express his or her opinion about whether we should go ahead with the sale. We went around the room. All but two of the ten people assembled supported going ahead. Many McCaw executives saw AT&T as a more secure future, and, despite its obvious bureaucracies, they were attracted to what looked like certainty.

I was one of two negative votes.

Craig did not reveal his own opinion about the sale but told us to go ahead and finalize the negotiations. I think he still hadn't made up his mind and was just gathering information, even then, at the eleventh hour.

The final price had not been decided, but Perry had helped set the stage by telling AT&T vice chair Alex Mandell a week or so before that the price had to begin with a "four" for McCaw—meaning at least $40 a share. Later on Friday, the price was set by the teams at the equivalent of $45 per share. At the time, the $12.6 billion purchase was the second most valuable in history.

Even then, Craig was not resolved. That evening, he asked me to go with him to meet with Ian Valiance, executive chairman of British Telecom, which owned 20 percent of McCaw's shares. We found Valiance in his suite at the Pierre Hotel. He was wearing a smoking jacket, had a fire burning in the suite's fireplace, and when we arrived he set aside the book he was reading. Craig told him that if BT made a higher counteroffer by morning, it could be McCaw's new owner. Valiance was friendly, but we heard nothing the next morning and proceeded to complete the negotiations with AT&T.

Still, Craig never considered a deal final until it was signed. So, to settle a few final points, Craig and I visited Allen in his suite at the Waldorf. Allen was watching a golf tournament on TV. Golf was a mandatory sport for all AT&T executives. But that did not dissuade Craig, who dickered with Allen on whether to add the value of the next quarterly dividend if the sale didn't close until after the dividend payment date.

The Sunday board meeting was in a small ballroom at the Waldorf. All of the directors were there, including Craig's brothers, Bruce and John, former Washington governor Dan Evans, and an assembly of lawyers and investment bankers. Although it was clear that a majority was prepared to approve the merger, the meeting was tense as I explained the details of the transaction. Evans, a board member for many years who had served three terms as Washington State governor and then went on to become a U.S. senator, spoke forcefully against the sale. McCaw still had lots of opportunity for growth, he felt, and would be swallowed up by the bureaucracy of AT&T. Seattle, meanwhile, would lose what he saw as an important engine of innovation. But when the final vote was taken, Evans supported the sale to make it unanimous.

In the end, Craig's decision to sell remains a mystery. He'd never much enjoyed the day-to-day operations of running the company, preferring to focus his energies on technology and strategy. And his efforts to find a successor among the ranks of former presidents and vice chairman had been frustrating. He did not relish the task of trying to rebuild the management team.

Although the agreements had been signed, the closing could not occur until we held a shareholder vote. Also, an antitrust lawsuit had been filed in Brooklyn federal district court to block the acquisition.

Craig quietly asked general counsel Roberta Katz and me whether there were legal grounds for McCaw to terminate the merger agreement. I told him that we could terminate if AT&T had made material misrepresentations to McCaw about their financial performance. Since their financial representations to us were the publicly disclosed financial statements, this would require us to challenge and prove that these were materially misleading. Craig said to go ahead and investigate. There was a potential argument that their financial statements were misleading because AT&T had a pattern every few years of declaring a substantial extraordinary write-down of assets. If this was part of a plan whereby each year they could report record earnings only to adjust them some years later, they would be engaging in misleading actions and reporting.

I consulted partners I knew at two premier law firms in New York. They said there might be a case but advised that it would lead to years of litigation and uncertainty for McCaw.

The draft merger agreement provided that Craig would join the AT&T board, but he asked that this be deleted. This was a shock to many, particularly at AT&T, since the CEO of an acquired company almost always joins the board of the acquiring company. But Craig didn't need the prestige, and he didn't want to be bound in any way by AT&T as he pursued future ventures, including competing against them. Besides, I think he considered attending board meetings when he was not in charge worse than going to the dentist. Craig also asked me to insist on deleting from the merger agreement a standard noncompete clause. After some resistance, AT&T relented, probably based on their big-company hubris.

The sale of McCaw closed in September 1994. Craig promptly retreated to his private offices on the top floor of an adjacent building and began competing with AT&T, including acquiring control of wireless carrier Nextel and investing in Clearwire. Neither achieved the success Craig had with McCaw.

Based on proxy disclosures, the value of Craig's AT&T stock was over $1 billion. Since his brothers had originally received equal shares of McCaw, *Forbes* ranked each as billionaires.

Barksdale was now in charge of AT&T Wireless, wholly owned by AT&T. Many of us had an unrealistic notion that our McCaw team could infect the hidebound telephone company with our entrepreneurial zeal, rather than having AT&T's bureaucracy crush us. It was not to be.[13]

A couple of months later in early 1995, Doerr recruited Barksdale, Peter Currie (our CFO), and Katz to become CEO, CFO, and GC of Netscape, which was to go public in September 1995. None of them had any background in the internet or technology expertise, but they were seasoned managers to balance tech geek Mark Andreessen.

I remained as president of LIN, now with 50 percent of it owned by AT&T and the remainder owned by the public. Under the shotgun agreement that AT&T had inherited, it was required before December 31, 1995, either to buy the remaining 50 percent at an appraised valuation or put 100 percent (including the shares they owned) up for sale. As much as AT&T valued McCaw's cellular markets, its largest was Seattle and a 50/50 joint venture with Pacific Tel of San Francisco. What AT&T really wanted were the four big markets owned by LIN: New York City, Los Angeles, Dallas, and Houston. But it would lose these if it didn't buy the other 50 percent of LIN. The price for those shares was to be determined by appraisers appointed by AT&T and the three independent directors.

I viewed my role as president as representing all shareholders by ensuring a fair appraisal process and maximizing the value of LIN. The three independent directors took the view that I represented the independent shareholders as they did. AT&T made it clear that they considered me an AT&T employee, and I should take orders from them. As the year progressed, and as AT&T took actions that I thought were an attempt to diminish the value of LIN, I increasingly sided with the independent shareholders.

Notwithstanding my independent stance, AT&T did not want to suffer possible legal claims from the shareholders if it removed me. Board meetings became increasingly contentious. At one such meeting, I counted fifteen investment bankers, lawyers, and AT&T staff in the room in addition to the nine board members.

To enhance the value of LIN, I pushed to spin off LIN's TV business that included TV stations in seven markets. This was successful and constituted an increase of $19.50 per LIN share.

AT&T's appraiser, Morgan Stanley, valued LIN at $105 per share ($5.6 billion) and the independent director's appraisers, Lehman Brothers and Bear Sterns, valued it at $155 per share ($8.26 billion). We finally reached an agreement on a valuation at $129.90 per share. I ordered coffee mugs imprinted with $129.90 + $19.50 = $149.40 and distributed them to the loyal LIN executives and directors.

We closed LIN's sale in September 1995. I was as eager to leave AT&T's employment as they were to be rid of me. Steve Hooper, a former McCaw executive and then president of AT&T Wireless, quickly and fairly negotiated a termination letter and I moved out of my McCaw/LIN/AT&T office six days after the closing. My plans included launching an investment firm with three friends and working on what proved to be a brief six-month adventure of bidding hundreds of millions of dollars provided by McCaw, Gates, George Soros, and others in an FCC-sponsored cellular spectrum auction. And I was planning on writing my check to invest in an internet bookstore.

Several of the McCaw alumni went on to become angel and venture investors, startup founders, and executives in Seattle's growing tech ecosystem. *Forbes* magazine labeled us the "McCaw Mafia" in a 2001 article.[14] The most notable is John Stanton, who joined McCaw in 1982, rose to be vice chairman and chief operating officer. After leaving McCaw in 1992, he created his own cellular empire consisting of Western Wireless and Voice Stream, ultimately selling it to Deutsche Telecom for $50.7 billion in 2001. It is now known as T-Mobile, the third-largest U.S. cellular company, with twelve thousand employees in the Seattle area. At least as successful as McCaw had been.

Craig publicly reappeared recently after being out of the spotlight for ten years with his SEC filing of a proposed public offering to sell $250 million of stock in a special purpose acquisition company (SPAC), Holicity, which could invest in technology, media, and telecommunications companies. SPACs are also known as blank-check companies because the investors give the company broad investment discretion. In one sense they are a public version of a private equity firm, but they have become popular as a way for a private tech company to become public by merging into the SPAC and to raise additional funding. Holicity later announced that it was acquiring a maker of small rockets used to send satellites into orbit, and Craig announced he was launching a second SPAC, Colicity.[15]

Recently Craig has also served as board chair of the Nature Conservancy and helped attract a $100 million donation from the Bezos Earth Fund to protect forests along the northwest coasts of Washington, British Columbia, and Alaska—areas frequented by Craig piloting his yachts and seaplanes for many years.

4

Microsoft and Amazon Innovate to Success

Microsoft and Amazon share the Seattle area but exist mostly in different business, social, and geographic spheres. The Seattle area, similar to the Bay Area's San Francisco and Silicon Valley, consists of Seattle and the Eastside, separated by twenty-two-mile-long Lake Washington, linked by two floating bridges.

Amazon is headquartered in downtown Seattle with most of its sixty thousand local employees, and Microsoft is in its suburban office park twenty-one miles away in the town of Redmond with its sixty thousand employees. Microsoft's campus consists of dozens of three- to five-story buildings of the type you would find in Silicon Valley, spread over a five-hundred-acre campus of green spaces, parking lots, and sports and cricket fields. The Amazon campus is multiple high-rise office buildings mixed in a dense urban environment of other tech and biotech companies, towering apartment complexes, restaurants, and shops.

Microsoft has no offices in downtown Seattle, but eight thousand of its employees live in Seattle and commute to the Eastside. Amazon, for most of its life, has been concentrated in downtown Seattle, but recently it has been acquiring and contracting to build high-rise buildings in the growing suburban city of Bellevue on the Eastside, with announced plans to grow its presence to twenty-five thousand employees. Both Amazon and Microsoft have operations and offices spread across the globe, but Amazon's remote software engineering offices are overall much more substantial.

The only comparable regional employer is Boeing, with more than seventy thousand employees, located adjacent to two nearby airports. Although employing an impressive number of workers, Boeing's regional importance has been declining for many years, along with the number of local employees, as it has moved many of its operations to California, the Midwest, and the South, with its headquarters now in Chicago. More dynamic and growing are the substantial local software offices of Facebook, Google, Apple, and many other global companies, together with significant homegrown tech and biotech companies, and a flood of startups, which combined employ another fifty thousand software engineers and others.

Microsoft was twenty-five years old when Amazon launched in 1995, and even then they were a powerful enough tech company that the Justice Department brought an antitrust suit against them a couple of years later.

Although both are software companies, they do not significantly compete on Amazon's e-commerce, digital media, and Alexa businesses nor Microsoft's office productivity tools, email (Outlook), operating systems (Windows), server software, or PC businesses. Both produce tablets, but Amazon focuses on its Kindle consumer reading and entertainment devices and Microsoft on its business Surface devices. Both failed in their entries into cell phone businesses, losing hundreds of millions (Amazon) and billions (Microsoft). They have a few areas where they cooperate, such as providing interoperability on their voice assistants' devices (Alexa and Cortana), although Amazon is the more significant provider of such devices.

The notable exception where they are fierce competitors is cloud computing, with Amazon having an estimated market share of 33 percent and Microsoft 18 percent, with the next closest competitor being Google at 9 percent. Amazon achieved an early lead when it launched its cloud services in 2006, and Microsoft (as well as Google, IBM, and others) did not significantly respond until seven years later, believing incorrectly that their enterprise customers would never substitute computing services from third parties for their internally controlled servers. Since Satya Nadella became CEO in 2014, Microsoft has increased its focus on cloud computing and appears to be gaining on Amazon.

In the future, we may see more competition between the two companies in artificial intelligence (which both use extensively), healthcare (where Amazon is making a larger commitment), and quantum computing (where Microsoft appears to have an early lead).

Microsoft

The Microsoft story begins at Lakeside School, an elite private high school in Seattle. In 1968 Bill Gates and Paul Allen became interested in computing by using an electromechanical teletyping machine that students at the school could use to log on to a remote mainframe computer.[1] Gates and Allen also began hanging out at the computer lab at the UW and using their mainframe computer until Allen received a letter from the university in 1971 asking them to stop using it.[2]

Coincidentally, at the same time, I was surreptitiously using the same UW mainframe to calculate which precincts should be prioritized based on past election results in the local political campaign for county prosecutor that I was comanaging.[3] My wife at that time, Mary, and I would arrive after hours and submit our punch cards to the student behind a counter (using Mary's account for the Physics Department where she was a PhD student). We would wait for thirty minutes or so until the computer printed out our results on long computer paper. Unlike Gates and Allen, we were not noticed, and our candidate won the election. They went on, however, to make a lot more money.

I was not aware of our overlapping use of UW computers, but at about the same time, in the 1970s, I learned from Bill Senior and Mary Gates that their young son Trey (as he was called by his family and their friends) was involved in something called "software." I tried to look up its meaning in the dictionary.

Microsoft's rise to be the dominant software company has been told in dozens of books and thousands of articles and, like Amazon, it is still a story nowhere near its last chapter.

By the mid-1990s, Microsoft's Windows platform was the world's leading software operating system, and Microsoft had triumphed over early productivity software companies by steady improvements and consolidation of its Word, Excel, and PowerPoint into a single Office package.

Gates and his leadership team were slow to recognize the emergence of the internet as a powerful way for people to connect and access information. Cable television was the dominant entertainment platform in the mid-1990s, and many industry leaders and observers believed that it would emerge as the way people accessed information, communicated, and bought goods and services from their homes.

When Craig McCaw and I visited Gates in his Microsoft office in 1994 as part of our discussions about a possible McCaw/Microsoft joint venture to deliver email, Gates pointed to a two-inch stack of documents on his desk that he said were agreements for a proposed joint venture with John Malone, CEO of TCI, a leading cable company. Gates said that even on his third reading he was finding clauses where Malone was "screwing" Microsoft. His deal with Malone was never finalized, but it illustrates the focus of Microsoft on cable as the network of the future.

In 1995, based on reports from younger Microsoft engineers of the growing excitement about the internet in universities and early business users, Gates sent his famous email to all Microsoft employees, directing them to infuse the Internet into all of Microsoft's products and services.[4]

Notwithstanding Gates's directive, Microsoft did not initially pursue the internet aggressively, partly due to the debilitating effect of the government's antitrust suit but more telling by a fierce internal battle between the leaders of Microsoft's traditional big-money-making Windows group (which had launched Windows 95) and their small internal internet group led by Brad Silverberg. Gates was surprisingly passive in the battle, and when CEO Steve Ballmer ruled in favor of the Windows group, Microsoft lost out in e-commerce (Amazon 1995), and social networking (Facebook 2004), and being the leader in search (Google 1999) and cloud computing (Amazon 2006).

Microsoft could have been a leader in all of these. It was only with Satya Nadella's election as CEO in 2014, replacing Ballmer, that Microsoft fully committed to the internet and cloud computing. Even though Microsoft's stock price plateaued during Ballmer's era, he rightly points out that during his tenure as CEO Microsoft made enormous profits, growing its annual revenues from $25 billion to $70 billion and increasing its net income 215 percent to $23 billion, derived mostly from the success of its Windows and Office groups—but it lost its innovative edge.

Today Microsoft sits atop the software industry with dominant platforms in operating systems and office productivity products. It stumbled badly trying to become a major mobile phone player with its disastrous acquisition of Nokia in 2013 for $7.6 billion. Microsoft's personal computers and tablets are strong offerings, even though they are not the leaders. They have now integrated the internet into their products and services, but Google's browser, search, and maps dominate them.

Nadella has quickly proven himself as an innovative leader, most prominently seen in Microsoft's aggressive development and sales of cloud computing products, providing a strong second challenger to Amazon. For the future, Microsoft has a leading position in quantum computing, and their HoloLens mixed reality glasses may become important.

Amazon

Bill Gates and Microsoft and Craig McCaw and McCaw Cellular would be plenty on which to build most cities into a tech powerhouse, but Seattle also has Jeff Bezos and Amazon, which has driven us into the position of being the second-most-important tech region in the world.

For twenty-three years, as a member of Amazon's board of directors, I have been involved in many of its major decision points. This includes a dinner meeting Jeff and I had with Barnes & Noble's Riggio brothers when they came to visit Seattle in 1996. They told us they would crush Amazon when they launched their website but added that they admired what Jeff had accomplished and would be happy to have a joint venture or even let Jeff run their site. He and I discussed it afterward, and then he called and politely turned them down. A bold decision for a young company but consistent with Jeff's oft-stated principle of not focusing on competitors but on Amazon's own business and its customers. In the week before our IPO in 1998, Barnes & Noble sued us on the grounds that our claim to be the world's biggest bookstore was misleading. It didn't stop our IPO and mostly gave our fledgling company more publicity.

I was involved in the decision to accept Kleiner Perkins' investment at a valuation of $60 million in 1996, the preparation for the IPO at a valuation of $429 million in 1997, and what was viewed, at least by outsiders, as the company's near-death experience during the dot-com bust of 2000. I was a part of the decision in Amazon's early days to reject a proposed joint venture with Starbucks because they wanted a large equity stake in Amazon in exchange for highlighting Amazon's books in their stores. I was part of the decision to launch Amazon Web Services (AWS) and later to extend it into serving the federal government. We all learned from these events, and Amazon grew stronger.

Following Amazon's successful IPO, it became clear to Jeff and the board that the public securities markets were willing to provide lots of capital for Amazon's growth and that losses would be tolerated. Even though Amazon's book business was growing, Jeff adopted a get-big-fast strategy by adding music and video disks, electronics, and home products. We made several large investments and acquisitions in internet companies, such as HomeGrocer.com and Drugstore.com at aggressive prices, most of which went out of business or were sold at losses.

We were able to borrow $1.25 billion in January 1999 by issuing convertible bonds, and Amazon's stock peaked at $106.69 a share. Jeff was named

Time magazine's Person of the Year in December 1999, which should have been a warning since this is sometimes a signal of bad times to come for the person.

Our stock price had begun to decline in 1999. As we worked in February 2000 to complete an offering of 690 million euros ($681 million) of convertible ten-year notes, the stock markets were about to take a dramatic turn for the worse. The NASDAQ peaked on February 28, 2000; after that, the world entered a severe recession.

Our executive team was in Europe meeting with prospective institutional investors as we worked to complete the offering. We learned from our executives that our underwriters were having great difficulty securing firm orders for the bonds and suggesting delay or cancellation of the offering. John Doerr, one of my fellow board members and a Kleiner Perkins partner, called me and said we needed to have a conference call with the lead bankers. We got on the phone with them, and Doerr made clear the importance to the banks' reputations of not backing out of the offering. Unsaid was that Amazon needed the money. The company was able to complete the sale of the notes, but the interest rate was 6 7/8 percent, significantly above the 4 3/4 percent Amazon had paid to sell convertible bonds in January 1999, and we had to agree to conversion terms that were unusually favorable to the purchasers. It was clear that no more money would be available for a while.

Just in time, the board and Jeff realized that the ship needed to be slowed. Jeff announced a get-the-crap-out strategy, which meant eliminating money-losing products and investments. We also reversed a management mistake we had made. In 1999 we had recruited Joe Galli—even though he had just accepted a job at PepsiCo that would have put him in line to be CEO—to provide Jeff with an experienced chief operating officer. Instead, Galli lobbied senior management and some of the board to replace Jeff as CEO. After thirteen months we asked Galli to resign.

Outside critics were predicting Amazon's demise. Amazon's stock hit a low of $5.97 on September 28, 2001. Amazon was able to reduce losses and even reported its first-ever quarterly profit in the fourth quarter of 2002—one cent per share. We survived. (Amazon's stock recently has been at over $3,000 per share—a nice return if you bought at the low. If you had bought $1,000 of stock at its low point it would now be worth at least $500,000.)

Lots of people ask me, "Why is Amazon so successful?" A few years ago, the CEO of a large European insurance company who was attending

Gates's annual CEO summit contacted me through a friend on his board and wanted to meet with me. When we met, he asked, "What are Amazon's secrets of success?" I told him, as I tell others, "Amazon has no secret management principles." Jeff talks about them all the time at "all-hands" meetings. He explains them in press interviews that can be viewed on the internet, and they are listed at the bottom of every press release.[5] But, I explained, you have to live by them all of the time, and most businesses are unwilling or unable to do so.

The most important is customer obsession. In his words, too many companies focus on their competitors and not on their customers. As Jeff explained most recently to a Congressional Committee, "Customers are always beautifully, wonderfully dissatisfied. A constant desire to delight customers drives us to constantly invent on their behalf."[6] I have seen Jeff many times make decisions that harm Amazon's bottom line but benefit the customer. As he says, tell me a customer who wants higher prices or slower delivery. Of course, in the long term, these decisions mean a bigger bottom line. A related principle is to think long-term.

The second principle is constant invention and innovation. As noted above, invention is closely linked to customer satisfaction. Constantly invent and apply technologies to solve problems and build new businesses. Customer satisfaction and innovation are powerful touchstones when making decisions. Decisions are far easier when you ask, "What is the best decision for the customer?" and "Is there a way to invent our way to a solution?"

The third principle is operational excellence. Examples of innovations in operations include two-pizza teams, one-click shopping, single-threaded leaders, working backward, and up-level hiring (where an employee outside the workgroup also interviews every applicant and has the power to veto their hiring).

Working backward is Amazon's internal process of beginning a project by writing the press release that would be used to announce a new service or product to the world. At most companies, the press release is the last step in launching a product, written by PR people, after the product and technology teams have created the new product. By starting with the press release, Amazon forces the team to focus on the benefits of the new product to the customer. Then the team must build the product or service to achieve those benefits. The draft press release is accompanied by FAQs answering questions that customers and the press might ask, such as the price, as well as describing technical challenges that need to be overcome.

Think long-term is the fourth touchstone, whether in launching new businesses or investing in new technologies. When businesses were just beginning to recognize the possibilities of machine learning and AI, Jeff told the board that he intended to use AI in every part of the business, and then he proceeded to hire AI experts and to train his existing engineers to use AI. Amazon created and made available to customers AI tools on AWS that they could use in their businesses—even to compete against Amazon.

Amazon also has fourteen leadership principles, some of which surprise people.[7] For example, "leaders are right, a lot" has more than a hint of hubris but is nuanced by commentary that leaders listen a lot and are quick to change their minds ("disconfirm their beliefs") based on differing evidence. "Have backbone, disagree and commit" is another compelling principle.

I have always thought it difficult to single out any one principle or process that is most important to Amazon's operations, but forced to choose, I suggest the importance of a single-threaded leader, particularly in launching new initiatives but also applicable to solving engineering and other problems.

An example of a single-threaded leader was Jeff's designation of Steve Kessel in 2004 to head up Amazon's effort to build a digital media business in books, music, and video, which ultimately involved building equipment like the Kindle, Amazon Fire, Fire TV, and creating content in video and books. Jeff could have asked the existing organization responsible for Amazon's physical books, music CDs, and videos—which amounted to 80 percent of their business at the time—to expand into virtual goods. However, Jeff recognized that giving authority to a single leader focused exclusively on digital media would more quickly create such a business without the distractions and compromises that might have occurred if the leadership was also running the physical business. This process was replicated with Andy Jassy's appointment to launch AWS.

Jeff has also publicly described his decision-making process, notably in categorizing decisions into one-way and two-way doors.[8] You can only go through a one-way door in one direction, you cannot undo your decision and come back. Examples are selling your company or quitting your job, although even for this there are examples of comebacks. Amazon's decision to offer free shipping for an annual prime fee was likely a one-way door. In two-way doors, you can later reverse your decision or at least terminate the action without enormously bad consequences. Building their mobile phone might have been considered a one-way door, but in actuality, it was dropped and Amazon moved on to building the Echo. One-way decisions require lots of thought and consideration; two-way decisions can be made quickly. It's OK to make mistakes—failures are learning opportunities at Amazon, as they are often considered by venture capitalists in the startup world. Jeff likes to say "we are experts at failure—because we have made so many of them."

In the venture world, if your company fails, it is not a black mark; you learn and can go on to start another company and receive funding. Of course, reasons for failures sometimes reflect a deeper problem with a leader. Jeff prefers speed, so he would rather have only 70 percent of the necessary information or belief and proceed with later course connections than wait until having certainty. Jeff, like Steve Jobs, never thought it valuable to do a market survey asking customers whether or not they would like some new product or feature when it has never existed in the past. Prime memberships at $79 and $499 mobile phones would not have been high on a list of customer's wants.[9]

Perhaps Jeff's overriding principle, which is not on Amazon's formal list, is his abiding optimism of the future and how we are only in Day 1. It is as infectious in his meetings with employees as his well-known laugh. His laugh is always near the surface and, although genuine, he subconsciously often uses it to mark some surprising insight or diffuse a tense moment.

At internal meetings, including board meetings, Jeff does not hesitate to use his powerful intellect or position to dominate a discussion of an issue, but he can also be a careful listener looking for new ideas. When I was the lead director, I once asked him what he hoped to get out of our board meetings. He replied, "If I get one good new idea out of every board meeting, it is a success." I thought it was a low standard and humbling. It also did not seem to represent the give and take at board meetings, which consist not only of management reports but also robust questions, observations, and suggestions by board members. And Jeff is not inhibited from challenging a point made by any of us. Occasionally his comment has a particular bite that signals deeper frustration or concern.

Board meetings span most of two or three days, including committee meetings and two dinners: some of the meetings consist only of board members with Jeff, others with Jeff and the S-team, and always a dinner with one of the S-team without Jeff present. Highly useful parts of every board meeting are private sessions of the board with Jeff. We would raise particular questions on our minds, and he would distribute a list of five important issues that were on his mind for discussion. Many of these issues were long-term or technology. They could include whether we were investing enough in some new technologies, such as AI or quantum computing. The private sessions could last an hour or more.

At Madrona, we have picked up Jeff's Day 1 philosophy to make the point that we want to help a company from the very beginning of the existence, Day 1. Jeff uses it in a broader sense, meaning that although the internet and Amazon may seem mature and in later phases to many, to Jeff we are still at the beginning. When some suggest, as I did once, that maybe Amazon is in day two, Jeff pounces: "No we still are in Day 1." It is his ultimate expression of optimism about what the future will bring. He wouldn't be investing $1 billion of his own money a year in Blue Origin if he didn't believe we are in Day 1.

Day 1 is also seen in his continuing willingness to launch new businesses, always innovating how to accomplish it and often creating new business models. The list is long and likely unrivaled in the history of American businesses: third-party sellers (1999, launched as zShops); AWS (2003); Prime (2005); Fulfillment by Amazon (2006); Kindle (2007); Direct Publishing (2007); Studios (2010); Fire Tablets (2011); Advertising (2012); Fire TV (2014); Twitch (2014, acquired); Echo (2015); Alexa (2015); Whole Foods (2017, acquired); Amazon Go (2018); and Amazon Care (2020). Amazon's new business launches were often greeted by skepticism from outsiders and sometimes insiders.

Many are platforms, meaning they are not only the smart use of a technology or creation of an app but are also a structure—physical, digital, or both—that others can also use and build on: retail (still the biggest), AWS, game platform Twitch (acquired the same year as the phone flop), and a new one will be healthcare. Somewhat longer-term, Amazon is planning to build and operate a global Wi-Fi network using 3,236 satellites launched by Jeff's Blue Origins.

Amazon has gradually expanded into financial services, loaning several billion dollars to third-party sellers, its store credit card for customers, and a Pay-by-Amazon for third-party merchants. The latter has not been widely adopted, at least partly because other retailers don't want to give Amazon access to a record of their customer's purchases. Transportation services, including electric and autonomous vehicles and drones, will likely be a major business offered to outsiders. Longer-term, any large industry that is susceptible to disruption may find Amazon asking itself whether it can make a difference.

Not all of Amazon's new business initiatives have been their technology inventions, but they have often combined new technologies and brought new business models to more established businesses, such as Amazon Go stores, same-day and one-hour delivery, transportation, fulfillment centers, and groceries. I believe one of the biggest initiatives will be their entry into healthcare. They are also one of the leaders in inventing and applying new technologies in the fields of robots (2012, with the acquisition of Kiva Systems), drones, and AI. I would love to discuss each of these in detail but here are my thoughts on a few to whet your appetite.

One of Amazon's most important innovations has been integrating third-party sellers into the Amazon retail marketplace on a par with sales of products owned by Amazon. This began as a concern expressed by S-team member Jeff Blackburn that eBay, with its successful auction model, might eat into Amazon's business. Notwithstanding Amazon's protestations that it ignores competitors, it took this concern seriously. Amazon acquired Live-Bid in 1996, a small online auction site founded by twenty-four-year-old entrepreneur, Matt Williams, backed by Madrona. But Amazon struggled to grow a significant auction business in the face of eBay's larger userbase. Williams then led efforts at Amazon to find a successful model for third-party sales and eventually integrated them with sales by Amazon, giving third-party sales an equal opportunity through its algorithms for "winning" the detail page, which is the landing page for customer product searches. This was initially considered heresy by some product leaders within Amazon. Today over 50 percent of sales are by third-party sellers, partly as a result of providing them services such as Fulfillment by Amazon (FBA) and operating loans so they can provide better customer service and grow their businesses.

Outsiders often marvel at the effectiveness of Amazon's webpages in selling products and its fulfillment centers in routing packages, but almost invisible behind the scenes are the thousands of individual processes and innovations that make this possible. Jeff Wilke, as the long-time head of Amazon Retail and Worldwide Operations, was responsible for conceptualizing and creating these systems, as he says, modeled on tried-and-true manufacturing principles and systems that he had acquired in his earlier years as an executive at Allied Signal. Who would have thought that over-the-hill manufacturing was a model for modern processing and delivery of e-commerce goods? Wilke did.

AWS grew out of Amazon's multiyear project to build a single software platform to replace the hundreds of systems and apps that had been built by groups throughout Amazon. These groups were often having to reinvent software processes. It was a mess. As they implemented the new platform, Amazonians working on it realized that they could create a similar software platform on its own servers where it could handle the computing and storage needs of other companies over fiber to eliminate the need to acquire servers and build basic software systems. Software giants IBM, Microsoft, and Google strongly believed that large businesses would never allow their important computer operations to be off company premises on computers owned and operated by a third party. At the time, their customers universally confirmed this. In a classic example of Clayton Christensen's theory of disruptive innovation, Amazon began with a small offering of compute services (Amazon Elastic Compute Cloud) aimed at startup businesses. Madrona hosted the first meeting of Amazon cloud staffers with customers in a loft on Seattle's Capitol Hill. To our surprise, over fifty young CEOs and software engineers showed up eager to sign up for a service that eliminated their need to invest large sums in computers and storage systems. They could rent by the minute or even second depending on their needs. Even as Amazon began to sell beyond small companies to bigger ones and invest in new hardware and many new services for AWS customers, its competitors sat on the sidelines. Jeff today marvels that they gave him a six-year lead. Jeff wishes he had received the same lead for Echo and Alexa, which also were initially panned by the technology press as, "Who needs it?"[10] But Google had learned a lesson and became a fast follower for intelligent voice devices.

Today, Amazon's share of the cloud market at over 30 percent is substantially more than Microsoft's (now a vigorous competitor), Google's, and Alibaba's, in that order. IBM and Oracle make a lot of noise but are far down the list of major cloud players.

One of many major decision points on AWS was a serious discussion by Jeff with the board as to whether we should pursue federal government business in light of the burdens of its bureaucracy and politics. There was no doubt the government needed help on its antiquated systems, and using the cloud was the fastest and least expensive way to do so. After our discussion, Jeff gave the green light on government business. When Amazon beat out IBM for the CIA contract, the wisdom of this decision was confirmed. Of course, the downside was more recently seen in the competition for the

Joint Enterprise Defense Infrastructure (JEDI) contract, won by Microsoft after vociferous opposition to Amazon's proposal by Oracle and President Trump. The result was litigation.

Amazon has not always been successful in entering new markets, as illustrated by its major failures to enter the China market and the mobile phone business. It also has pulled out of other business efforts that were deemed not sufficiently promising, including event tickets, restaurant deliveries, and discount coupons. During its annual budgeting exercise, Amazon prioritizes investments and is willing to eliminate new proposals to allocate funds to the most promising.

Jeff often personally invests a large amount of his time in big new projects, such as Kindle and Echo. For the phone, he moved his family to the Bay Area for several months so he could work daily at their subsidiary Lab126 where they were developing the phone.[11] It was launched in an uncharacteristically splashy event in a theater in the Fremont District of Seattle, where Google's local offices are located. We all cheered. But quickly it became apparent that the phone's major invention, an intricate way to use multiple sensors to display 3D images, was not compelling.

The phone was a flop. Instead of following a Christensen strategy of a low-priced entry, it was a high-end costly phone trying to mimic Apple's success. Some had argued internally that Amazon's innovation should be an integration of cellular with Wi-Fi networks it could license and expand to improve and lower the cost of service as its entering strategy. Amazon's phone was also greatly hampered by Google's insistence that to license its Android operating system and access its app store, Amazon would have to use Google's major apps such as Google Maps and Google Message as the default apps. Although Google's position has been criticized as a possible antitrust violation, Amazon chose to fork an earlier version of Android that did not give Amazon access to Google's app store. Having to build and induce outside developers to build apps, Amazon couldn't bridge the gap in apps and was at a huge disadvantage. To its credit, Amazon recognized that it was not going to succeed and took a $170 million write-down and exited the business. It was an Amazon failure, although some viewed the Echo, which was launched one year later and also developed at Lab126, as the natural successor. The Echo was announced with a press release and no splashy event.

One of Jeff's strengths is that if entry into a new business begins to make traction, he is not afraid to pour money and top talent from throughout the company into the project, even if it means losing money for many years. Billions were invested in building AWS, although it relatively quickly became very profitable. Billions of dollars and top talent have also been invested in groceries, Echo/Alexa, and India with long runways to profitability.

Amazon has a long history of trying to sell groceries successfully over the internet. In the late 1990s, it was a 20 percent minority investor in HomeGrocer along with Madrona and Kleiner Perkins. It went public in 2000, but after a disastrous merger with its rival WebVan, the combined companies folded. Although many customers loved the service, it could not achieve sufficient density of deliveries and size of orders to become profitable.

Years later an internal memo prepared by one of the executive teams in connection with Amazon's annual budget process argued that to continue to grow its retail business in the future, Amazon would need to be a major player in the grocery category, pointing out that groceries capture the largest share of the consumer's wallet and are a weekly and sometimes daily purchase. In 2007, Amazon launched its grocery delivery service, Amazon Fresh. After trying many different business models, by the late 2010s, it still had not significantly penetrated the market and was unprofitable. It decided to increase its efforts with physical stores and faster delivery. It launched Amazon Go stores, small convenience stores with the innovative feature that cameras and sensors record the customer's actions so that the customer can select items, put them in a bag, and walk out of the store without stopping at a checkout clerk. A bill is automatically sent to the customer electronically. Amazingly, it is quite accurate—justifying Jeff's insistence that sufficient accuracy could be achieved with vision and AI.

Amazon acquired Whole Foods in 2017, one of the smaller grocery chains that emphasize healthy foods. Amazon then embarked on a separate grocery effort, building grocery stores from the ground up, experimenting with several different sizes and implementing innovations that it tested in Go Stores like automatic checkout and palm print ID. With same-day delivery, including from Whole Foods and other stores it opens, and with the added impact of COVID-19 causing increased online shopping of groceries and other products, Amazon will likely succeed in building a large grocery business. Although it will be a technology innovator, it will not be dominant given the large number of competitive grocery outlets, including much larger grocer retailer Walmart.

Amazon's attempt to establish a significant retail presence in China after many years of effort and hundreds of millions of dollars has also been a failure. Finally recognizing the unlikelihood of success, Amazon substantially cut back its effort. This contrasts with India, where Amazon is making huge investments with significant losses. It has become one of the two leading online retailers in India. Amazon learned lessons in China, one of which is that it place in charge a very talented leader. India's leader is Amit Agarwal, who had earlier been Jeff's technical assistant—a job that many promising employees have held, such as Andy Jassy, CEO of AWS.

It has long been accepted business school dogma that companies should not enter new businesses that are significantly different than their existing businesses. For years GE successfully violated this dogma, building railroad engines and refrigerators, as well as a substantial finance business. But it eventually experienced financial difficulties and had to unwind its conglomerate empire. In 1995, GE was the second most valuable company in the world. Today it is not in the top thirty.

Amazon is the exception to this rule. Think of its different businesses: physical goods, digital services (AWS), video production. The only thing that seems to tie them together is Amazon's core principles: customer service, operational excellence, long-term thinking, and invention. Contrary to business theory, Jeff appears not to be concerned that Amazon does not have specific expertise in a new business or cannot leverage its other businesses (although this is sometimes true). He does apply at least two of the three criteria that he used when he decided to acquire the *Washington Post*: Is it an industry that needs to be disrupted, and can Amazon make a difference? His third criteria for the *Washington Post* was whether it was important enough for him to become involved. This certainly applies to his Blue Origin space venture. And maybe it applies to all of Amazon's businesses in the sense of improving people's lives by delivering goods and services conveniently and at a low cost.

Jeff sometimes asks about whether or not a category is a land rush: If Amazon doesn't enter now, will it be too late to enter after others have acquired a dominant position? On the other hand, Amazon entered the video streaming business after Netflix had established a commanding presence, and Jeff has said there is plenty of room for both Netflix and Amazon. Now with multiple strong streaming competitors such as Disney, Amazon continues in a strong position, since it provides its streaming offering as part of Prime with lots of other benefits to the customer that others can't match, such as free delivery of physical goods.

To continue to grow its revenues, Amazon needs very large markets to go after, and I expect Amazon to grow healthcare into one of their largest business platforms, on a par with retail and AWS. Bigness in business has become a target of politicians, public policy groups, and the press. Certainly, all big institutions, including government and business, should in Jeff's words expect and deserve scrutiny. At the same time, large businesses have the scale and resources necessary in some industries to best drive innovation and change for everyone's benefit.

Healthcare is a huge $3.8 trillion market, buried in paperwork and tied up in politics. Amazon's initial efforts in healthcare were mostly to improve the delivery of healthcare services to its employees as well as innovations in diagnostics and the treatment of diseases. Amazon's entry will eliminate inefficiencies and lower costs, bringing together its separate healthcare initiatives of the last few years as it builds another Amazon platform. Healthcare not only needs disruption, but innovation on a large scale can also make a difference—something Amazon excels at.

Amazon launched Amazon Care in February 2020 as a primary care service for its fifty-three thousand Seattle-area employees and their families. It combined telemedicine checkups and in-person at-home or in-office doctor's visits and prescription deliveries. Amazon pays for the service and saves on healthcare costs. "The service could also shrink health care costs for the company, a concern for employers nationally as premiums rise at nearly twice the rate of inflation, to an average of roughly $20,000 per family in 2019, according to the Kaiser Family Foundation."[12] By early 2021, Amazon had opened seventeen neighborhood health centers for employees and their families in the areas around Dallas–Fort Worth, Detroit, Louisville, Phoenix, and San Bernardino.

～

Not unexpectedly, once Amazon had launched Amazon Care to its own workforce, Amazon announced in early 2021 that it was beginning to offer the service to other companies for their employees. Eventually we can expect Amazon Care to be available to individual customers.[13]

Broadening its healthcare portfolio, Amazon acquired PillPack in 2019 for roughly $753 million, an online pharmacy that ships prepackaged prescription medicines to people's homes. Its introduction of Echo Buds, its version of earbuds, could well evolve into hearing aids—even though subject to restrictive regulation.

Amazon further plunged into consumer health care in August 2020 with its launch of Halo, an AI-powered health and wellness service using a new wristband that measures body composition, tone of voice, sleep, and activity tracking. Halo's goal is to help you improve how you sleep, move, sound, and feel by discovering, adopting, and maintaining personalized wellness habits. Research suggests that information watches and rings may measure changes in heart rates and other vital signs that signal the onset of COVID-19 and other diseases before there are any external symptoms.

In addition to improving the delivery of healthcare, Amazon is investigating how to better discover cures for diseases. For example, Amazon researchers are applying machine learning to cancer research in partnership with the Fred Hutchinson Cancer Research Center (the Hutch), which is a brief walk from Amazon's headquarters in Seattle. The initiative uses machine learning to analyze unstructured patient records, doctor's notes, and other data to identify the best outcomes for various treatments.[14] Amazon's response to COVID-19 has accelerated its efforts and ambitions in healthcare services, diagnostics, and cures.

Amazon's ambitions do not stop with healthcare. Investments in building its delivery network to supplement and possibly replace services it uses from UPS and USPS have expanded to include investments in electric and autonomous vehicle companies and research. As difficult as it would have been a couple of years ago to imagine Amazon becoming an automobile manufacturing company, its investment of $700 million in Rivian, a startup developing electric trucks, and then its acquisition of Zook, a developer of autonomous vehicle systems for $1.2 billion, put it squarely in the business of building electric and autonomous vehicles. What started as building a delivery service and vehicles for its own use can eventually morph into businesses serving others. These are certainly large businesses and likely to be disrupted.

Jeff told the board on several occasions that his greatest fear was the board would select as his successor a CEO who is not willing enough to take big risks. Fortunately, with Andy Jassy succeeding Jeff, Amazon will continue to be led by a creative risk-taker. At this point, large new markets and large risks are necessary for Amazon to grow. And there are lots of big industries that need to be disrupted. With Jassy as its new CEO, expect to continue to be surprised by Amazon.

Part 2

5

On the Precipice of the Future

Five technology trends—wireless connectivity, internet, AI, quantum computing, and biotech—will change everything about our lives, from healthcare to transportation.

We are assaulted daily by technologies unimaginable fifty years ago. Teleputers in our pockets can communicate with anyone in the world and instantly access the greatest trove of information ever assembled. The benefits to us are magnificent, but we worry about the problems they cause. The internet of things and the flood of smart devices create conveniences but put more of our private lives on the internet from what we eat (smart refrigerators) to our physical condition (Peloton). Our iPhones allow us to connect twenty-four hours a day, but they track us anytime anywhere. Facebook allows us to connect and share with friends, but our information is mined to bombard us with ads, and use by young people can intensify depression.

During good economic times, we tend to forget the benefits of these technologies and instead focus on issues they raise. Politicians of both parties and many individuals and their lobby groups often argue that the benefits are outweighed by real and imagined privacy concerns. Some demand moratoriums on the use of new technologies such as facial recognition in crime-fighting. The challenge for society is to develop policies, laws, and technologies that can keep and expand the good uses and prevent the bad ones.

Many in tech cities such as Seattle and San Francisco choose to ignore that tech companies provide the economic engine that has brought prosperity and opportunity to its residents. They look upon the companies more as a source of revenue than as contributors to solutions to the cities' social issues. For example, they adopt laws that protect the taxi industry and inefficient portions of public transportation by taxing mobile car services, even though these taxes raise prices on usage by low-income groups. Even when an economic and health crisis hits us, such as the COVID-19 pandemic, too many public and private leaders are unwilling to advocate that the benefits of cell phone tracking outweigh privacy concerns.

COVID-19 upended economic and technology forecasters who already had notoriously poor forecasting records. Nonetheless, as a venture capitalist, part of my job is to identify new technologies that can be important to people and businesses and then to invest in new applications of those technologies.

The big trends appear obvious to me and are in accord with the views of most other technologists: five technology sectors—ubiquitous wireless connectivity, the internet, artificial intelligence, quantum computing, and biotech/healthcare—will transform the world over the next twenty years. I also believe they will transform our cities, even though many are skeptical of this view.

These technologies are not separate silos and, as I will discuss, they often depend on one another and are made more powerful by working together. Other technology innovations, such as robotics, commercialization of outer space, and virtual and mixed reality will also be important, but their timing and impact on our lives are less certain. Will augmented and virtual reality, once predicted to be a major way we communicate, learn, and access knowledge, realize these predictions?

As I mentioned earlier, when predicting the future, I have benefited from several technologists who were unusually farsighted in grasping the power of new technology and placing a big bet on it, including Craig McCaw, George Gilder, Jeff Bezos, Craig Mundie, my partners and associates at Madrona, and the many entrepreneurs and inventors who have flowed through our Madrona offices over the past twenty-five years.

Ideas about computers can be traced back to Charles Babbage who might be said to have invented the first computer in the mid-1850s, although he did not build one. If you define a computer as a calculating machine, then the first computers were the abacus and eventually the slide rule (which I carried to college).

The modern computer age began in the 1940s with computers using vacuum tubes, punched paper tape, and punch cards as the input/output connections—costing millions of dollars each. Computers and software advanced through the 1950s and 1960s in government and university labs and private companies, with innovations involving separate memories in digital computers to store instructions and data. This was based in part on theoretical work by John Von Neumann and Claude Shannon and the development of the first high-level software language in 1953 by Grace Hopper, which evolved into COBOL and a compiler for translating the high-level language understandable by *humans* into the low level instructions to control the machine. Equally important was the introduction and improvements of microprocessors by Intel and others.

In the 1970s two college dropouts from San Jose and two from Seattle began playing around with personal computers, started a revolution in computing, and became business giants. Steve Jobs and Steve Wozniak focused on designing a personal computer, and Bill Gates and Paul Allen on writing the programming tools and operating software. As competitors, but each needing the other, they built two companies that came to dominate the PC industry, so that today Apple and Microsoft are two of the three most valuable companies in the world. Their hometowns, the Bay Area and the Seattle area, have the number one and two tech economies.

In the 1970s and 1980s, a diverse group of mostly California university faculty and graduate students, with support from the Advanced Research Projects Agency (ARPA; now DARPA), developed the wiring, switches, communications protocols, and software for the fundamentals of digital networks and **inter**connected **net**works that made the internet possible. A critical innovation was by Tim Berners-Lee, a researcher at CERN in Switzerland, who released a paper in 1990 describing a HyperText Markup Language (HTML) that could be used by different web browsers to render text, images, and other material, resulting in the World Wide Web.

The purchase of McCaw Cellular and LIN Broadcasting by AT&T in 1994 and 1995 for $18 billion marked the coming of age for mobile phones by marrying the traditional phone company with the developers and operators of the first national cellular network. The year 1995 also marked the commercialization and consumerization of the internet with the IPO of Netscape highlighting their graphical internet browser, Gates's famous memo to all Microsoft's workforce declaring that all Microsoft services and products should embrace the internet, and the launch of Amazon's e-commerce website.

In addition to wireless mobility and the internet, computing has also been transformed by cloud computing, first promoted by Amazon in 2006. Previously conducted by machines on company premises, Amazon provides on-demand computing and storage services remotely through high-speed connections. Initially only of interest to small tech companies who realized the benefits of renting computing and storage by the hour, minute, and second, today big companies are rapidly outsourcing their computing to the clouds provided by Amazon, Microsoft, Google, and others.

If the computer, mainframe and personal, was the most important technology of the last half of the twentieth century, and the internet, mobile teleputing, and cloud computing were the most important additional technologies of the first twenty years of the twenty-first century, my nominations for the most important additions over the next twenty years are AI and biotech, along with the new applications they will make possible, such as autonomous vehicles, robotics, and cures for many diseases. AI is becoming ubiquitous and increasingly powerful. The breakthroughs that are beginning to occur in healthcare by the unraveling of the genome and the proteome and the use of AI are leading to the understanding of human biology by powerful AI, and eventually the ability to design and implement changes to our biological system using techniques like CRSPR, synthetic proteins, protein folding-AI, and other molecular mechanisms.

We also are on the cusp of Bezos's Blue Origin and Musk's SpaceX replacing the U.S. government as the conquerors of space. Quantum computing is on the horizon, and VR/AR/MR will improve with miniaturization, more computing power, and improved mobility and applications. And in the brains of thousands of innovators across the globe are blockbuster inventions that will surprise us.

Artificial Intelligence

Microsoft CEO Satya Nadella, speaking at our 2020 annual meeting of Madrona investors, declared that AI is "the defining technology of our generation."

AI is already being integrated into many applications, including personalized e-commerce recommendations, lower shipping costs, optimizing the operation of fish farms, video interpretation, and improved medical diagnostics. Soon, AI will help discover new cures for diseases and make autonomous driving widespread.

Artificial intelligence has been on our minds for years. In popular media, we learned about HAL in 1968 and R2D2 in 1977. Even earlier Isaac Asimov introduced us to smart robots and pronounced three laws to keep them from harming us.[1] My first brush with artificial intelligence was back in 1988 when I sent a letter to my son John, then an eighteen-year-old sophomore at Williams College, with a *New York Times* article about "fuzzy logic" tucked into the envelope. A precursor to artificial intelligence, fuzzy logic used mathematical formulas to represent vagueness (hence the term *fuzzy*) to quantify decision making based on imprecise information. John was majoring in math, so I suggested that he keep an eye on this arena. Today John runs a fund that uses AI exclusively when making investment decisions—that is, predictions—about which stocks to purchase and which to sell.

In popular culture, artificial intelligence is software that can do many of the things the human brain can do, such as image recognition and problem solving. More technically it is algorithms that are able "to correctly interpret external data, to learn from such data, and to use those learnings to achieve specific goals and tasks through flexible adaptation."[2] Machine learning and deep learning are subparts of artificial intelligence.

Many scientists have contributed to the development of AI during the past fifty years, but machine learning and deep learning did not begin to be widely used in applications until the 2010s. Since then, its use has spread quickly among academic researchers and business engineers, increasingly playing a role in business and consumer applications.

We have witnessed the evolution of AI as it has mastered checkers, chess, the Rubik's Cube, Go, AlphaGo, and poker to defeat the best human masters. Initially, it learned how to play by being taught strategies and being fed the best games of past masters. More recently it is not taught but learns on its own by being given only the rules of the game and developing winning strategies by playing millions of games against itself.

AI makes e-commerce more efficient and lowers shipping costs. It augments photo and video interpretation—scanning airport crowds to find suspected terrorists, for example—to increase public safety. It improves medical diagnostics. In the future, AI will help us discover new cures for disease, and it is at the heart of advances in autonomous driving.

Clearview has created an archive of billions of photos of people from the internet and, when fed a photo of someone, can find all the matches in its database and show where the person has visited. It is not yet foolproof, but as it acts it learns and improves. It is also controversial.

ML is being quickly applied to a wide range of business processes and decision making. Amazon, in Bezos's words in 2016, was using AI in "demand forecasting, product search ranking, product and deals recommendations, merchandising placements, fraud detection, translations, and much more."[3]

Machine and deep learning are enhancing existing apps and making many new intelligent apps possible. The Uber mobile app, taking advantage of geolocation and mobile payment systems, has now been enhanced by AI to improve the routing of vehicles. Intelligent video combines AI, low power, high-fidelity video, smart apps, and connectivity through the internet. These systems are being used by governments, private companies, and individuals: video cameras installed by law enforcement, businesses, and individuals in private and public places for crime deterrence and detection; authorizing entrance to buildings and stadiums; increasing security at airports; and acting as invisible cashiers at Amazon Go stores.

AI is proving valuable in health care by analyzing medical images, helping diagnose diseases, and inventing new proteins—but will it be able to diagnose and recommend treatments of individuals better than any doctor or doctor and AI working together? Many in the medical profession accept that machine learning can be better than humans at working through all the research papers, human trials, and vast amounts of data to suggest diagnosis and treatment, but they stop short of accepting that the doctor should not be the final decider. Craig Mundie has his answer to the question: "Human biology is too complicated for humans. If it is too complicated for humans, is it too complicated for machines?" He answers no.[4]

The term *artificial general intelligence* (AGI) is mostly used to describe advanced forms of AI that more closely mimic the human brain in solving problems from a base of general information without being trained for a particular domain. AGI can learn and integrate knowledge from multiple domains, learn incrementally, and reason across those domains. By comparison, AI is limited to solving problems in specific domains where the software has been provided with a relevant database and the software has been trained for that problem. A major challenge for AI is to develop software that can "learn how the world works by watching video, listening to audio and reading text."[5]

Microsoft is supporting the development of AGI by investing $1 billion in OpenAI, headed by Sam Altman, whose goal is to develop software and the requisite hardware capable of doing anything the human brain can do.[6] In 2020, in an important step in its quest for AGI, OpenAI released GPT-3, a language model using deep learning that can perform a broad range of tasks, like summarizing text, writing news articles, translating, and answering SAT questions. It can generate articles with only a few human prompts that are often difficult to distinguish from human authors. It uses

175 billion parameters trained on a general corpus of text from the web, which includes about a trillion words.[7]

Even beyond AGI is the possibility that AI could truly mimic the human brain with consciousness, self-awareness, feeling, and the ability to initiate independent thought and action.[8] In the view of Elon Musk, the late Stephen Hawkins, and others, it is only a matter of time before we build sufficiently powerful computers with access to enough data that they will equal and then exceed the human brain.[9] Many leading software and other technologists believe that when computers become smarter than humans, with the ability to initiate strategies, they will be existential threats to humans. Ray Kurzweil calls the moment "singularity," which he predicts will occur in about 2045.[10]

Whether this will ever be achievable is hotly debated. While conceding that machines, "from a 3-D printer to a plow," can outperform a human, Gilder challenges the "eschatologists" of Silicon Valley by arguing that superfast computers do not lead to "imagination, creativity and independence." He argues in *Life After Google* (2018) and other writings that "the blind spot of AI is that consciousness does not emerge from thought, it is the source of it." He finds support in the Gödel-Turing difficulty of self-reference and quantum uncertainty.[11]

I also like the whimsical suggestion of a leading technologist that, if computers become smarter than humans, why do we fear that they will take over the world when they are more likely to choose to spend all their time playing computer games? But the serious point is that we are making computers smarter and smarter, and we don't know for sure where this leads.

Rather than waiting for machines to bypass humans, Craig Mundie raises the question of why we are not taking into consideration our ability to reengineer the human body and mind to "hybridize" ourselves and our descendants, combining the power of AGI and biotools like CRSPR.[12] This raises challenging social and moral issues of whether we are prepared not only to cure diseases and enhance our capabilities but to create a "mixed-biology" species – a mixture of carbon-based and non-carbon-based intelligence for super-humans. Walter Isaacson in his new book, *The Code Breaker*, raises some of the obvious issues of bioengineering to improve our species, but this is only the beginning of our society's debates.[13]

Likely, the most noticeable impact of AI on people's lives in the next ten years will be when the use of autonomous vehicles becomes widespread. With improved AI, we are on the technology path to achieve this.

Ten years ago, in our search at Madrona for cutting-edge technologies in which to invest, we invited a well-known consulting firm to make a presentation about the viability of driverless vehicles. We thought driverless vehicles might revolutionize urban mass transit—a not-incidental concern in Seattle, where some commuters spend a total of 138 hours a year sitting in traffic—and we wondered if it might be smart to invest in these technologies.[14] We envisioned fleets of autonomous shared-ride vehicles that could be summoned via mobile phones, shuttling workers to driverless public transit, and easing our overcrowded highways while cutting the number of accidents due to human error. We imagined blocks of parking lots transformed into public green spaces—a cityscape transformed. But the consultants convincingly told us that we should not expect autonomous vehicles in our lifetime, if ever. They predicted that the sensor and software technology would never be good enough, governments would not allow it, and people would not accept it. So we put further investigation on the shelf.

Then a couple of years later, Tesla launched its car with early autonomous features, and two of my partners bought Teslas. We also discovered that two of our young portfolio companies had developed technologies that were being sought by major auto companies developing their autonomous vehicles. Echodyne had developed a new type of radar that they thought would be useful for large drones, but, in addition, all major auto companies wanted to test it. Another of our startups, Mighty AI, was using crowdsourcing on the internet to tag data sets for machine learning algorithms, such as whether a picture of a sweater was crew neck, V-neck, or turtleneck. It became obvious that, more importantly, they could tag whether it was a person or a traffic sign in a street scene—of obvious interest to anyone working on software for self-driving cars.

Autonomous driving has progressed more slowly than expected, but machine learning combined with a variety of sensors, including video cameras, radar, and lidar, is constantly improving it. Accuracy can be enhanced by location signals from other vehicles and street signals. We are already seeing early pilot projects on public roads such as Waymo's driverless taxi service in a suburb of Phoenix and autonomous trucks being tested on interstate highways, and Tesla autos essentially piloting themselves on major highways. Even with the timetable of adoption uncertain, we can expect that the effects—environmental and economic—will be profound.

We are in stage one of aerial drones being flown by remote operators, with some autonomous features such as collision avoidance. Stage two is already possible with totally autonomous flights.

Biotech/Healthcare

After decades of working to unlock the secrets of our genes and proteins, researchers are making significant progress in cracking the codes and manipulating them with CRSPR, synthetic protein folding, and mRNA. They also are using the body's immune system with immunotherapy so that breakthrough cures for cancer are finally arriving. There is optimism regarding the development over the next ten years of many more medicines for curing a variety of diseases. There also are promising efforts to improve the delivery of healthcare through the use of technology, even though we still face big challenges because of a bureaucratic, regulatory, and insurance-encrusted system. These efforts are driven by nonprofit research institutions, innovative hospital systems, and the entry of innovative tech companies such as Amazon that provide hope for improving the quality of healthcare and lowering the cost for everyone.

The recent coronavirus pandemic has added new urgency to finding cures and preventive vaccines for diseases and viruses and improving the delivery of healthcare to people. COVID-19 spurred a wave of global cooperation among scientists, institutions, and governments, with the collateral benefit of an uptick in investments for the discovery of new drugs and delivery-system improvements. We will be able to more quickly detect and contain the next virus; develop effective treatments; and create, test, and authorize vaccines.

In 1981, after the U.S. Supreme Court held that genetically engineered bacterium could be patented, startup companies such as Genentech and Biogen quickly became successful in San Francisco with medicines for treating a limited number of deadly diseases, such as multiple sclerosis.

At the same time, as discussed in chapter 3, scientists Steven Gillis and Christopher Henney—both working at Seattle's Fred Hutchinson Cancer Research Institute (the Hutch)—founded Immunex to bring to market a drug based on Interleukin 2 (IL-2), a protein that activates killer T cells, that was showing promise in fighting cancer. But demonstrating the difficulty of cracking the codes and overcoming all the ways that cancer continues multiplying, notwithstanding attempts to destroy it, Immunex's IL-2 drug killed too many patients. It was not until almost forty years later in 2017 that Phil Greenberg and Larry Corey, researchers also at the Hutch, were able to develop a process using IL-2 killer T cells to cure cancers while controlling deadly side effects. They helped revitalize the field of immunotherapy and spun off a company from the Hutch, Juno, that went public in

2015 and was acquired by Celgene in 2018 for $9 billion. The initial research at the Hutch that led to Juno was funded partially by charitable contributions from Jackie and Mike Bezos, Jeff's parents. After meeting with Greenberg in his lab, I alerted Jeff to the promise of Juno. He then invested in Juno's first round of financing. I was fortunate to have been involved with both companies—as the lawyer and investor with Immunex and an early investor in Juno.

In addition to gene splicing and immunotherapy, another fruitful area of biotech is proteomics, which Mundie predicts will be important in detecting and curing cancers and other diseases.[15] Proteomics, the study of proteins, has the potential for helping provide personalized medicine specific to each person's particular needs arising from their unique combination of genes and proteins.[16]

Proteins serve crucial functions in essentially all biological processes. Since they circulate in our blood stream, a snapshot of proteins in the blood is a proxy for the state of all our body systems. Using AI to understand these mappings of protein concentrations allows scientists to determine our past and present conditions and to predict to a significant degree what our likely future biological state will be. Because of this, small samples of proteins contained in blood can indicate cancer and other diseases.

Several companies are developing ways to detect cancer at an early stage by analyzing proteins in the bloodstream, which would greatly improve our ability to identify cancer and other diseases, including targeting cures with newly created synthetic proteins or other biological structures that bind to proteins. These include SomaLogic in Boulder, targeting a broad range of diseases, where Mundie is an investor and an advisor; Grail in Menlo Park, California, whose founder and CEO, Hans Bishop, was also a cofounder and CEO of Juno; and Nautilus Biotechnology, which was founded in 2018 in Seattle by Sujal Patel, who had launched and built a software storage company, Isilon Systems, in 2002.[17] Madrona was a founding investor in Isilon Systems, which was sold to EMC for $2.25 billion in 2010. Madrona also is now an investor in Nautilus, which is creating a platform for analyzing and quantifying the human proteome. Even though Patel is a software rather than biotech expert, he is similar to many software engineers, such as Mundie, who have a keen interest in biotech and recognize the opportunity. These companies also are a reflection of the competitive nature of biotech, with multiple companies seeking to develop competitive services, diagnostics, and drugs.

At the Institute for Protein Design at the University of Washington, David Baker is a prolific researcher and inspiring leader who has built a lab consisting of over fifty graduate students, postdocs, fellows, and research scientists, as well as a large alumni body that is focused on creating de novo designed proteins (synthetic proteins) to solve medical and technology problems.[18] Baker Labs has spun off several companies based on their protein research, and Baker recently asked me to join him on a three-person nonprofit board set up by Bryan White, often an investor in these spin-offs, to disburse proceeds from cash and stock contributions from Baker, White, and others. In response to COVID-19, Baker and his labs are using their protein expertise and tools to develop a COVID-19 therapeutic, vaccine, and inexpensive test kit.

In addition to Baker Labs' work on protein folding, Google's AI off-shoot DeepMind appears to have solved one of biology's biggest challenges by determining a protein's 3D shape from its amino-acid sequence, possibly the most important biotech development since CRISPR.[19]

In July 2020, Madrona invested in the startup Pine Trees, which included a Seattle-based founder and CEO and researchers from MIT, developing a fast, inexpensive test for COVID-19, based in part on research by Feng Zhang, one of the inventors of CRISPR technologies. At the same time, my eighteen-year-old grandson, Charlie Anderson, was a lead researcher on a high school team that was developing an inexpensive COVID-19 test that could be used by labs in the developing world.

Baker, Greenberg, and Corey are not unique in creating labs of researchers and spinning off companies. Researchers at universities and hospitals around the world are forming new companies, licensing technologies from their institutions, and securing much of the funding from angels and venture funds. Most innovations in biotechnology are coming not from Big Pharma but university and medical research labs funded by the National Institutes of Health and private donations.[20]

Pharma companies are well equipped to conduct trials and market new drugs. They often license the innovations from these new companies and ultimately buy the companies, sometimes after they have gone public. This provides financial rewards for the inventors, labs, and employees, which is reinvested into the local economy and back into further biotech innovation. An example is that after Juno was sold, investors in Juno reinvested in Grail, another spinoff of the Hutch, recruited former Juno CEO Hans Bishop to be CEO, and hired former Juno employees. A flywheel.

Biotech startups generally require large amounts of capital. Bob Nelsen, a venture capitalist at Arch Ventures in Seattle and the Bay Area, is the master of this, having led the funding of Juno, Grail, Sana, and other start-ups, raising hundreds of millions and billions of dollars. Nelsen seemingly is always on the move and on his phone. Madrona subleased office space to Nelsen for over twenty years, but he was seldom in his offices. He would often breeze into my office, pulling his suitcase that I assumed was packed more with documents than clothes, with his mobile phone to his ear, telling me that he had a breakthrough new company and that I had to talk with them. Our conversations were often punctuated with 15-second interruptions on his phone. Not all his suggestions became successes, such as his recommendation that I invest in a company turning algae into jet fuel, but one of those references led me to meet Greenberg, the founder of Juno.

Nelsen raised over $1.5 billion for Grail in its first two years from Illumina, Arch, Sequoia Capital, Gates, Bezos, and strategic pharma and institutional investors. He also led the financing for Sana, announced in 2020, raising $821 million from Arch Ventures, institutional investors, and Bezos. Creating new pharmaceuticals has always been expensive, but Nelson and his financing skills have raised the art of funding to a new level.

Although much progress is being made by thousands of researchers in using new biotech learnings and tools to discover new disease-curing therapies, the underlying science of our genes, mRNA, and proteins is devilishly complicated. This was recently brought home to me in conversations with Rob Bradley, a researcher at the Hutch. Originally trained as a mathematician with a physics degree from Princeton, he went on to earn a PhD in biophysics from the University of California at Berkeley. One of the areas that his Bradley Lab is working on is unraveling the mechanisms by which mRNA can become corrupted as it converts DNA into proteins, thereby causing cancer. He has developed high-speed techniques for analyzing thousands of assays to identify mutations in mRNA related to the production of cancer cells that can lead to eliminating these mutations by revisions in the genetic code.

Notwithstanding challenges, the future is bright for biotech innovation in San Francisco, Boston, New York, and Seattle—all cities with leading medical research labs working with cutting-edge technologies of immunotherapy, gene editing, and proteomics.

Even if cities do not have a history of biotech startups, business leaders and investors in every community with a university medical school or hospital system with researchers should be prioritizing investment in their local institutions. If combined with entrepreneurs and angel or venture funding, it can help get their healthcare flywheel moving.

The entry of information technology companies into biotech and healthcare is further accelerating the expansion and improvement of healthcare services. I described Amazon's extensive initiatives to develop healthcare services in chapter 4. Alphabet, Apple, and Microsoft are also moving ahead with their efforts to improve the delivery of healthcare at lower costs and better service by launching medical devices, apps, and services.[21] Alphabet is analyzing billions of searches to predict the spread of diseases and is building an AI system to predict dangerous medical conditions in individuals, such as kidney disease, to give doctors an early warning and consider whether to recommend dialysis.[22] Apple has included ways to monitor health conditions on its iPhone and Watch.[23] Microsoft is leveraging its cloud services and AI skills in conjunction with established healthcare systems and startups which see the potential of blending Microsoft's software skills with their biological prowess without having to do it all themselves. Startups are developing apps for early and less costly ways to diagnose diseases. For example, a startup is developing an AI-driven system that views the retina at the back of each eyeball to identify signs of chronic illnesses and conditions such as diabetes, hypertension, arteriosclerosis, optic nerve disease, high myopia, and age-related macular degeneration.[24]

Amazon, however, has launched the most comprehensive effort to enter the healthcare field using its retail network, AWS, and innovations in AI.

Quantum Computing

Quantum computing will be one of the fundamental inventions of the twenty-first century, allowing us to use computers to accomplish things we have never been able to do. Quantum computers will complete complicated calculations in seconds or minutes that would require hundreds of years to complete on our present digital computers. This is because they make use of quantum theory, which instead of having each 0 or 1 in a calculation represent the unique binary digits, 0 and 1, calculate using "qubits" that can represent ones, zeros, and "superpositions" of ones and zeros.[25]

Researchers and companies believe that quantum computing's initial impact may be most important in simulating quantum mechanics in chemistry and materials science to design new molecules, chemical reactions, catalysts, and materials.[26] Building a quantum computer is extremely difficult, involving physics, electronics, and software. The Chinese government is spending large amounts to develop a quantum computer, which overshadows the amount being spent by the U.S. government, and some worry that this will provide China a major advantage of being the first to develop a quantum computer; however, since U.S. government quantum programs are mostly classified, we don't know if this in true.[27]

For commercial applications, Microsoft, Google, and IBM have large quantum programs, investing billions of dollars to develop a quantum computer that may more than match Chinese programs. These companies have entered into partnerships with leading quantum researchers at universities around the world, thereby locking them up for their projects. For example, Microsoft has formed a consortium to develop a scalable topological quantum computer with TU Deft in the Netherlands, University of Sydney, Copenhagen University, University of California at Santa Barbara, and Purdue University. They are working on a specific novel theoretical approach (topologically protected qubits) computer to overcome one of the problems of quantum computers when "qubits interact with their environment and lose their quantum information before computations can be completed."[28] Microsoft has built a quantum computer simulator on Azure that allows developers to learn about quantum algorithm development and to experiment with a variety of different systems made from different types of physical qubits while all these quantum hardware systems are still in their infancy.

Amazon, which had not invested in quantum on the same scale as other companies, has recently been recruiting quantum experts to build its quantum computer.[29] In 2019, Amazon also launched an AWS service that allows scientists to explore and experiment with quantum computing hardware being developed by other companies, initially limited to partnerships with some smaller companies. They also announced a research partnership with professors at the University of California at Berkeley.[30]

Quantum computers were first envisioned by the physicist Richard Feynman in 1982. He was trying to do computer simulations of quantum mechanical systems and discovered that the complexity grew so quickly that it would never be possible to do on a classical computer. So he conceived of a way to use quantum mechanical computation methods to do that simulation. It wasn't possible to build a quantum computer then, but the theory was interesting. In 1992, Peter Shor was the first to show that a mathematical computation, in this case factoring large prime numbers, which also was computationally intractable on classical computers, could also be solved if one had a quantum computer. Since factoring large primes forms the basis of most modern day cryptographical systems, this drew lots of interest. This potential threat and opportunity motivated a number of government actors to invest in research and development in this area over the years. Because this was more understood than the potential benefits of the quantum systems simulations, it because the poster child for early work both in practical system development and in exploring other applications. Accordingly, the most widely discussed potential use of quantum computers has been to crack the security codes used by most of the world's banks and financial institutions. These codes rely on finding the unique factors of very large prime numbers which can only be determined by existing computers in hundreds or more years. A sufficiently large quantum computer will be able to find these factors for a single prime number in the range of a few days to hours. To avoid this problem, many are working to develop new encryption methods using algorithms that are based on other mathematical constructs that cannot be cracked quickly by quantum computers.[31]

Space

In late May 2020, SpaceX launched two astronauts to the International Space Station. It was the first time since 2011 that the United States did not have to use Russian launch and space vehicles for sending astronauts into space. Until recently, governments have provided most of the leadership and funding for exploration of space through agencies such as NASA and contracts with commercial companies such as Boeing and Raytheon. But now private companies, conceived and funded by wealthy tech entrepreneurs, together with government contracts, have stepped to the fore providing new ambitions and achievements.

The leaders are Musk's SpaceX and Bezos's Blue Origin, but many smaller companies are also pursuing aerospace efforts. Bezos is spending over $1 billion of his own money annually on Blue Origin and is focused on creating lower-cost launch capabilities for reaching space. Bezos views this as the necessary infrastructure for successfully conquering space. Rob Myerson, the former CEO of Blue Origin, discussed his views with me over lunch of the current trajectory of space commercialization, which he believes will include, by the middle of the decade, commercial space stations and commercial transport to these stations and the surface of the moon.[32] Bezos's dreams include building bases on the moon and people living in space and on the moon; Musk's dream is building cities on Mars.

Too early was Craig McCaw's Teledesic, which proposed in 1994 to launch a system of 840 satellites (an audacious number at that time) into low Earth orbit to provide "internet in the sky" service, solving latency and user capacity problems that limit the number of simultaneous users of higher geosynchronous satellites. Craig signed me up to help achieve his dream and appointed me to the Teledesic board. He raised hundreds of millions of dollars from Gates and Saudi investors, but the satellite and antenna technologies were not yet ready, and the effort eventually was disbanded.[33] The effort did create a cadre of satellite engineers who increased Seattle's expertise for the next generations of space systems.

Now the development of small satellites and improved antenna systems is making launching thousands of satellites for a satellite-to-ground Wi-Fi and perhaps phone communications possible. The United Nations estimates that "almost 4 billion people around the globe" are underserved for internet access.[34] Musk's SpaceX is launching hundreds of satellites into low Earth orbit for its Starlink broadband data constellation, which is projected to grow to nearly twelve thousand satellites in low Earth orbit. SpaceX's facility in Redmond, Washington, is playing the lead development role.

Amazon is developing its system using 3,236 satellites for low Earth orbit to "provide low-latency, high-speed broadband connectivity serving tens of millions of people who lack basic access to broadband internet."[35] Their satellites will be launched by Bezos's Blue Origin, which is headquartered in the area south of Seattle where Boeing Aerospace developed Boeing's space programs for many years before departing for warmer climates.

Madrona has identified at least fifteen private companies in the Seattle area working on space projects, ranging from rocket rideshare services for launching packages into orbit on other company's launch vehicles (Spaceflight Industries) to a company focused on "the moon, and beyond" (Aerojet Rocketdyne).

At lunch a few years ago on top of Seattle's Space Needle, I sat next to Prince Guillaume, the hereditary grand duke of Luxembourg, discussing his country's $26 million investment in Seattle's Planetary Resources. The company planned to mine asteroids—a good idea but woefully premature. Madrona declined to invest.

On the Horizon

We are only beginning to understand the full impacts of robotics, virtual/augmented/mixed reality, distributed ledger systems, and 3D printing. The question now is whether we will have "a robot for every person," just as prior innovators imagined computers on every desk and teleputers in every pocket.

Professor Luis Ceze's lab at the University of Washington is developing techniques to store data in DNA.[36] At the current exponential growth of the world's digital data and its storage, a gap will develop between our storage needs for data, such as medical images and places to store it. DNA storage will be able to store digital information in a space that is orders of magnitude smaller than data centers today and will be more permanent. Ceze, a part-time Madrona venture partner, recently launched OctoML, a UW spinoff that aims to make machine learning more efficient, secure, and accessible for a wide range of engineers and hardware.

The invention of metamaterials in the 1990s is what Nathan Myhrvold calls "the closest thing to magic he's ever seen."[37] It can be used in a wide range of applications from flat-panel radars for autonomous vehicles to antennas for satellite communications.[38]

Since the first days of the internet, we have had to type to access its power, but now we are entering an era of multisensory interfaces of voice, touch, and video. Amazon Go stores use video to recognize your purchases as you pick from shelves so you can walk out of the store without human interaction. At home, you can talk to your Alexa device to find out the weather, ask for music to play, or get help with homework. Typing by humans may become as archaic as handwriting. AI combined with voice and video will provide you with a very smart personal assistant.

In the next few years, with 5G networks and service from thousands of satellites, we will have true global connectivity from mountaintops to farms.

Cities that encourage the application of tech to help solve their transportation, education, public safety, and healthcare needs will be the ones most likely to set new flywheels spinning and attract creative people for building economies of the future.

All these and other unforeseen inventions will play big roles in creating new tech winners. Autonomous vehicles will substantially replace manually driven vehicles on our roads, restricting the latter to driving only certain roads off-hours. Our skies will be filled with autonomous drones carrying freight and maybe people. AI will replace doctors for many purposes. Improvements in AI and its application to every aspect of our lives, from transportation to healthcare, will transform our lives—I believe for the better.

6

Investing in the Future
Talent and Capital

My early investments were in small amounts. As an angel I usually invested in companies I was involved with as a lawyer or because I knew and trusted someone involved. These include Advanced Technology Laboratories (ATL), Immunex, Seattle Silicon, Advanced Digital Information Corporation (ADIC), Active Voice, VISIO, and a half dozen others. Sometimes I lost my investment, but enough were winners that it seemed reasonable to invest in local technology startups.

I mentioned earlier how Gordon Kuenster as CEO had guided ATL to a successful outcome for investors. Kuenster was a true entrepreneur, always wanting to bet on cutting-edge technology and willing to take a risk. But most of his endeavors after ATL were unsuccessful, and I invested in all of them. They include Seattle Silicon (automating the design of chips) and Virtual Vision (a 1990s attempt at Google Glass), a Paul Allen–backed company. But I did not begrudge Kuenster.

All my early investments were as a part-time angel. It was not until 1995, when I resigned from McCaw and LIN, that I colaunched Madrona and became a full-time investor. Venture capital is a business, and it is about money, but I have always believed it is also part of the creative arts. Creative individuals come to us with dreams to build something meaningful and with the drive to do so. Success is visibly measured in dollars—but at its core, for the best companies, it involves not software but people. It is about trust, shared successes and failures, creating comradeship, and working to make a better world. It is never fun to lose money on an investment, but hopefully something good resulted for the people involved or the ecosystem.

I remember being an investor and part-time developer in the 1980s of a real estate project in Seattle's Pike Place Market. At one point, we had a hole in the ground and no bank willing to finance our condo/retail project with ten low-income units. At the time, I said that I would be happy to lose my money if our project could be built. Fortunately, we changed the project to a small hotel, Inn at the Market, with the ten low-income units, received a loan, and made money.

All of Amazon's first million dollars in 1995 were raised from angels (and Jeff's family). At the time, Amazon did not appear to be an obviously better investment than any of the hundreds of other angel-financed ventures in the 1980s and 1990s. The investors were local individuals, often friends of friends, the founder was new to town and untested, and the business plan highly speculative.

So how do you determine what will be a good investment? I have described earlier some of the lessons that can be drawn from my investment in Amazon, which I think are relevant even if you don't count the unexpectedly stratospheric upside. The broad principles are: invest in emerging technology trends; support smart, dedicated founders; back innovative products or services that people can benefit from; accept valuations even if they push the envelope; don't be scared away by traditional competitors; and be willing to take a risk.

As an angel, I didn't perform the extent of deep due diligence that Madrona performs today on all our prospective investments, which becomes particularly important when investing other people's money. But even with rigorous due diligence, most early-stage investments come down to an informed hunch.

Investors sometimes argue whether the technology or the founder is the most important. The easy answer is both. An innovative product or service is obviously important, but our experience is that it is very difficult to create a successful company without a great founder. Often a great founder can adjust their product to find a market fit to make it successful. You don't have to love the founder, but it adds to the pleasure of the investment. In my experience, investing in unpleasant founders often leads to unhappy outcomes because such people usually cannot recruit and motivate great employee teams.

Most successful founders believe they are changing the world for the better, and this vision is what drives them forward—often when others think it's wrongheaded. But extreme egos, where a founder refuses to consider the advice of others, is a road to disaster. One of Amazon's bedrock principles is that leaders listen to alternate views and be willing to change their minds.

Beyond a strong founder, what's essential to a business's success is providing a product or service that people will pay for. It sounds elementary, but when I first meet with entrepreneurs, they too often begin presentations with a recital of the enormous potential market. Sometimes they cast their product or service as one that they need to sell to only a small percentage of a very large market, as if that proves how easy it will be to cash in. (If, for example, the total potential market is $10 billion, then 1 percent of that market suggests sales of $100 million—in their logic, a financeable company.) After about ten minutes of this, I can't help interrupting: "Why don't you describe your service and talk about why someone will pay money for it."

The exception to this rule is advertising-based companies, like Facebook or Google, that offer a free service but attract so many users that advertisers will pay hefty fees to reach them. Notwithstanding the success of these two giants, this is a risky premise on which to build because it requires an enormous number of users to make meaningful revenue from advertising. For example, ten million site visits per month may sound like a lot, but if the advertising revenue is $0.10 per visit, this totals only $1 million a month, whereas a successful venture-backed company usually should have revenues of at least $50 million per year.

Bezos's original assertion that Amazon could sell lots and lots of books over the internet, for millions of dollars a year, was unprovable at the time.[1] While I could see that he was smart, when we first met I did not recognize Bezos as all that different from the best of other founders. He had a degree from Princeton and lots of personal charisma, but only after years of witnessing his accomplishments do I feel confident observing that even among very bright people, Bezos is unusual in his ability to apply his intelligence to solve complex problems. The strongest fact supporting my investment in Amazon was growing evidence that lots of people were going to use the internet. The most important negative was that even if people started buying books over the internet, Barnes & Noble, the dominant book retailer, would jump in with its vaster resources and dominate internet sales—just as it had retail stores. This is the kind of argument often made as to why a startup won't be able to challenge established companies.

But I already knew a David could succeed against a Goliath. I'd seen it during my ATL days when that ultrasound innovator launched its business in an industry dominated by giants like Siemens. Or when upstart Immunex created new drugs in an industry run by Big Pharma. Even my experience at McCaw Cellular confirmed this principle. Our young firm had challenged the Bell Operating Companies so successfully that we ultimately sold to AT&T for $16 billion. All of this is to say that startups have some inherent and important advantages—namely, nimbleness. They can innovate much more quickly than large companies, and they don't have legacy businesses that would be undercut by a new model. (In his 1997 book, *The Innovator's Dilemma*, Clayton Christensen evangelized this view.)

I have followed these lessons throughout my investment career, often losing money. But my gains significantly outweigh my losses. Risk is a necessary corollary. A few years after investing in Amazon, I led my venture capital firm to invest in HomeGrocer.com. Customers loved the service. The venture firm Kleiner Perkins joined us in investing, and HomeGrocer .com went public. But after a disastrous merger with Webvan (which I had strenuously opposed), the combined companies went bankrupt. The investment decision wasn't necessarily bad—for if you never invest in a Home-Grocer.com, you'll never invest in an Amazon. It should be easy to tell the difference, though it seldom is. That is why at least one-third of venture investments fail.

My Amazon investment also points to an important lesson: a founder is sometimes unable to raise enough capital to adequately launch his or her business. I'd minimized that risk by telling Bezos that I wanted to invest but only when he had raised his full $1 million in startup dollars. I was indeed taking a chance by imposing this condition. Bezos could have raised his million and closed the door to me. But four months after our conversation, he kept his word and called to say he had the full amount, so I could now invest. Rather than being offended, Bezos recounts this story as an example of a prudent tactic that impressed him.

Though venture capital investments have become a larger source of funding for startups, angels continue to be important as the early risk capital and precursors to more substantial venture financing. For business leaders in cities looking to build their tech economies, developing a local ecosystem of angel investors focused on startups is critical in getting the flywheel moving.

In 1995, when Bezos struggled to raise his first $1 million, Seattle did not have a robust funding ecosystem—particularly when compared with

Boston and Silicon Valley. We had just a few active angels writing checks, a handful of venture firms with modest capital to invest, and small investment teams. Developing the healthy funding infrastructure that allowed us to become one of the world's leading technology ecosystems was a step-by-step process that took more than a decade. I offer it here as a model for other cities to build upon.

In 1995, a group of us who were interested in developing a tech economy in Seattle visited Silicon Valley to get a better sense of what they were doing. As we visited with angels and venture capitalists, young biotech CEOs, and leaders at Stanford, we were shocked to learn how far behind we were compared with their number and engagement of angels and large venture firms in supporting Bay Area startups. Of course, we knew that the valley had a history of successful technology firms going public or being acquired—starting with Hewlett Packard and Intel's predecessors. But still, the depth of their technology ecosystem was startling. Hewlett, Intel, and Cisco, among other firms, had created numerous wealthy executives and investors, who were in turn investing in startups. Thus, Jim Clark, who had cofounded Silicon Graphics, and the venture firm Kleiner Perkins invested in and helped nurture Netscape before it went public. On that visit, we saw the valley's flywheel at work very clearly: success begat wealth, which led to increased investment, which spurred new firms, and begat more success.

We came back to Seattle determined to build a tech startup ecosystem and, in 1997, formed the Technology Alliance to promote the technology industry and legislation such as support for the UW's engineering department and increasing broadband to K–12 schools statewide. Bill Gates Sr. was our first chair. We launched the Alliance of Angels (AOA), with the express purpose of helping educate potential angel investors and increasing the availability of early-stage capital. The AOA held monthly lunch meetings where startup founders pitched investments in their companies to our assembled member angels. Some angels had struck it big (or small) at previous startups. Some were wealthy former executives or investors from Microsoft and later Amazon. Over the twenty years since, the AOA has grown to more than 150 members investing a collective $100 million in about two hundred startups. So far the AOA has celebrated more than forty successful exits through company sales or IPOs. In 2015, when the AOA launched a $5 million seed fund, Madrona and Steve Singh, cofounder of Concur, were lead investors. Singh later became a managing director at Madrona.

Our Madrona firm began in 1995 as a quartet of friends and colleagues—Bill Ruckelshaus, Jerry Grinstein, Paul Goodrich, and I—who had each made some money in the worlds of tech, business, and corporate law. At Madrona, we would investigate potential investments and make presentations to the others. Each of us would decide, individually, whether to invest, writing a check to Madrona that our firm would then invest.

Ruckelshaus and Grinstein were close friends and former *Fortune* 500 CEOs at Browning Ferris and Burlington Railroad, respectively. Goodrich, a partner of mine at Perkins Coie, had been most recently a partner with a Chicago venture firm focused on environmental companies. After Ruckelshaus left government in 1985, I recruited him to join Perkins Coie as counsel. Other than Goodrich, we had never worked at an investment firm. Also, as luck would have it, all of us were out of work.

The internet was coming to the fore as a potentially powerful business and consumer platform, cellular technology was maturing, and tech startups were erupting across Silicon Valley and, to a lesser degree, in Seattle. All in all, a propitious time for a group of eager new investors.

We often met with entrepreneurs at the local Starbucks in Seattle's Madison Park, which is CEO Howard Schultz's home Starbucks because he lives nearby. One of the young members of our Madrona team, Keith Grinstein, Jerry's son (now deceased), held so many meetings there that his friends installed a plaque in his honor on one of the tables.

During Madrona's early years, each of us decided whether to invest and how much, with our joint contributions ranging from $50,000 to $1 million. Often our investments were in conjunction with other local angels and sometimes with Silicon Valley venture firms we invited to join the financing. Many of these young companies went out of business with no financial return to us. But over our first four years, nine of our companies went public and six were sold for gains, with a total increase in value of six times our investments (an annual IRR of 35 percent)—not counting Amazon.

Partly luck, partly tech boom times. But we also benefited from our careers in business before becoming angel investors. As Paula Reynolds, former CEO of Safeco and current board member of GE and BP among other global companies generously wrote of us recently: "It must be remembered they had lifetimes of experience before they came into venture investing. They were an overnight sensation 30 years in the making."[2]

The work involved a lot more than making financial bets on a company's future success. Many times we also helped companies get up and running. For example, Greg Gottesman, whom we hired in 1997 fresh out of Harvard, wrote the business plan for Nordstrom.com in our conference room, sitting across the table from Dan Nordstrom. I brought in Bill Gurley at Benchmark Capital, a leading valley venture firm, to coinvest with us. The parent company, Nordstrom, owned 80 percent of the new business, and when the expected IPO did not happen, Nordstrom bought us out at an originally negotiated return of three times our investment.

Today, twenty-three years after we made our initial scouting trip to Silicon Valley, Seattle has hundreds of angel investors. Trilogy Equity Partners, led by former McCaw Cellular executive John Stanton, who went on to found T-Mobile, is the leading firm of professional angels. Stanton and his partners differ from a venture firm only in that they invest only their own money, with no outside investors. Additionally, Bezos (Expedition Ventures), Gates (BGI), and the late Paul Allen (Vulcan) have their own investment offices that invest in local and national early-stage tech companies. For example, Madrona has coinvested with Gates in Echodyne, Bezos in Pro and Nautilus, Trilogy in seven companies, and in numerous companies with Vulcan.

Beyond wealthy individuals, corporations like Amazon and Microsoft are also active tech startup investors—locally as well as globally. Even Boeing, which never had much of a startup culture, has revived its mothballed investment program to support early-stage investments in electric planes and aerospace technologies. Some corporate investment programs make money—but if that is their goal, it is a mistake. The primary purpose of corporate investment programs should be to have a window into the most recent cutting-edge technologies and business models. Cities interested in generating a tech-economy flywheel should encourage locally headquartered companies, whether tech or not, to create startup investment programs.

The first step, however, for cities to get their flywheel spinning is developing an ecosystem of individual investors willing to support tech startups. Banks and traditional financial firms are not likely sources of early-stage risk capital. And the small size of initial angel investments rarely justifies the time and expense for an out-of-town venture firm to perform the necessary due diligence. For those reasons, developing a local angel investment community is essential.

When developing an angel community, potential angels should be made aware of the special tax incentives for investing in startups. Taxation is not the principal reason angels and venture firms invest in startup companies, but tax incentives can encourage wider participation among investors and encourage them to value a long-term focus for their investments.

In 2010 I was appointed by President Obama to the National Council on Innovation and Entrepreneurship (NACIE) to make recommendations on changes in federal laws to encourage more investments in early-stage companies, in recognition of the fact that a substantial percentage of new jobs in America are created by high-growth startups. The council was staffed by the Department of Commerce and encouraged by secretary of commerce Gary Locke, former governor of Washington State. Under the able leadership of cochair Steve Case, the billionaire cofounder of AOL, we recommended strengthening the tax incentives to encourage long-term investments in small businesses and providing a new path for very small companies through crowdfunding to raise less than $1 million without normal SEC requirements.[3]

Fortunately, we were able to work directly with the talented staff of the Obama White House. Following a meeting in the Roosevelt Room with Valerie Jarett and other key presidential advisors, President Obama agreed to support our recommendations in Congress. The legislation provided that 100 percent of capital gains taxes are waived on direct equity investments in operating companies with gross assets of less than $50 million at the time of the investment if the investment is held for at least five years.

Months later, as several of us stood in the Rose Garden watching the president sign the legislation, we were somewhat amazed that it had passed Congress. Several liberal members of the Senate, which was controlled by the Democrats, opposed the legislation as did the New York Times in an editorial,[4] but President Obama's recommendations pushed it through to adoption.

In addition to local angel investors, a limited amount of capital is available from a few national sources that should not be overlooked. A leading example is Steve Case's Rise of the Rest Seed Fund. Headquartered in Washington, DC, the fund has invested in more than 130 startups in over seventy cities. Case also suggests that smaller cities encourage the largest corporations in town to form locally focused venture funds. This has happened in Columbus, Ohio, where Nationwide, State Automobile Mutual Insurance, and Grange Insurance have each started venture funds.

When I visited Oklahoma City, I met with a local fund, i2E, Inc., that has been able to attract investments from numerous out-of-state venture funds and angels into companies that the fund identifies and, in effect, is the local representative of the other investors for managing the investments.

Case contrasts this boot-strapping approach with the frenzied contest among two hundred cities competing against one another to win Amazon's favor as the corporation's new headquarters. "Take half the energy and half the capital you are willing to devote to Amazon and put it toward your startup sector," Case said. "That will bear far greater fruit over the next ten to twenty years." A study by his firm identified ten smaller urban centers—including Cleveland, Indianapolis, and Portland—as likely places to invest, not least because the cost of doing business there is up to 57 percent lower than in San Francisco. An added bonus: with less investment capital focused on these places, valuations are lower. Case also cites St. Louis as fertile ground for startup incubation because of its comparatively low cost of living, nearby universities, and the fact that several new venture capital funds have launched there since 2013.[5]

If a local startup begins to gain noticeable revenue tractions, traditional venture firms from around the country will happily become investors in these later-stage deals. It is the initial $1 to $10 million of investments that are the most difficult and almost always have to come from the local community.

Each city is different, and each will need to assess its strengths to determine how best to get the flywheel moving. Beyond angels, venture firms, and corporate investment arms, other models can aid startup creation, such as business incubators, accelerators, and boot camps. I discovered a number of these in the Midwest cities I visited, and many cities have their own local versions. They deserve local support.

In Seattle, we have several incubators, accelerators, and workspaces where creative people rent space, network, and share ideas with other like-minded entrepreneurs. Sometimes companies are launched out of these group workspaces. Incubators may hire a dedicated group who work together to conceive and launch startups—doing everything from researching market potential for a new product to hiring a small team, creating a website, and building the code for a prototype. Some incubators have investment capital—often obtained from outside angel investors and even venture funds.

A leading example is Pioneer Square Labs (PSL) in Seattle, which was founded by Greg Gottesman, a former managing director at Madrona. A startup studio that describes itself: "We develop ideas with entrepreneurs from scratch and validate or kill them (ruthlessly)."[6] PSL has launched twenty-five companies and killed over 100 ideas in its five years. Madrona was a founding investor.

Many of the companies created from these models are unsuccessful, but they contribute to the tech ecosystem by stimulating investment ideas and providing experience to entrepreneurs. The occasional successes and training by experiences of entrepreneurs justify those efforts.

One of the early successful incubators was begun by Brad Feld in 2006 in Boulder, Colorado, initially investing between $6,000 and $18,000 in early-stage companies and providing an intensive three-month mentorship. Techstars, as it is known, has expanded to spawn Techstars accelerators in Boston, Seattle, New York City, Austin, and internationally. Each year the Seattle Techstars receives over a hundred applications proposing ideas for new companies from entrepreneurs all over the world. Of these, ten are selected for a three-month program where their teams work in a common space at the University of Washington, receiving guidance from a small staff and outside mentors. Each team receives $120,000 for living and startup expenses and free in-kind services such as AWS credits. In exchange, Techstars gets a 6 percent common stock interest. Madrona helped found Techstars Seattle and has been an investor/sponsor ever since. At the end of the three months, Techstars sponsors a pitch session for the ten companies in front of local venture firms and angels. Most receive startup funding, averaging $1 million. Madrona has invested in several. Since 2010, Techstars Seattle has graduated about one hundred companies that have raised over $1 billion.

Accelerators in Silicon Valley have mentored thousands of startups, with hundreds of successful exits. The best known and arguably most successful of these is Y Combinator, in Mountainview, California, whose successful participants include Airbnb, Dropbox, Stripe, Reddit, and Twitch.[7] At Madrona Venture Labs, our in-house incubator, a dedicated team works with entrepreneurs and recruits founders to launch new companies.

Another kind of studio, the Riveter, founded by Amy Nelson, provides membership-based collaborative workspaces "by women for everyone."[8] Madrona was an initial investor in this model, which aims to promote networking among startups, investors, and financial and small business experts. It has workspaces in Seattle and LA and is developing a virtual model spurred by the impact of COVID-19.

Like Stanford in Silicon Valley, an important contributor to Seattle's tech ecosystem is the Department of Computer Science and Engineering (renamed the Allen School of Computer Science and Engineering in 2018) at the UW, which can be viewed as a superincubator. From Madrona's founding in 1995, we have been strong supporters and partners working closely with many of the faculty, including particularly Ed Lazowska, who was chair of the school from 1993 to 2001. Ed creatively led the faculty into a new era of innovation and entrepreneurship to match their research and teaching. He also led efforts to create strong relationships with the tech business community.

Recognizing their importance to our tech community, I cochaired the UW's campaign in 2010 to build a new computer science building that would increase faculty and add student collaboration classrooms and labs around a four-story central atrium. The campaign raised $80 million from individuals and businesses, and it enabled the department to double the size of its faculty and number of annual graduates. Previously, the UW had focused most of its fundraising on UW alums and big companies. I suggested that we focus the campaign on tech leaders, most of who had graduated from other universities, as I had. This was viewed with skepticism by the then UW president, but by emphasizing the importance of the engineering and computer science department in the building of a strong tech business community, more than a majority of dollars donated to the campaign came from nonalumni.

Even earlier, in the 1980s, local tech leaders had organized the Washington Research Foundation to provide funding to faculty who were developing technologies that appeared to have commercial viability. A number of us contributed to its formation, and it became the beneficiary of substantial royalties under a biotech patent. Over the years it has provided significant funding of UW startups, and Madrona often has coinvested with them.

Madrona holds monthly office hours in the atrium of the Allen School, where we meet with grad students and faculty to talk about their work. For more than fifteen years we have awarded the Madrona prize to the students whose projects best combine cutting-edge research with commercial promise. Madrona has invested in twenty companies founded by the Allen School's faculty, and in 2015 we helped raise funds for a second new building that again allowed the Allen School to double its faculty and student graduates. The founders of several of our successful companies were major contributors.

One of the Allen School's leading professors, Oren Etzioni, founded the Allen Institute for Artificial Intelligence (AI2), funded by $100 million from Paul Allen. Etzioni has been a part-time venture partner with Madrona for over twenty years and the founder of three Madrona-backed companies. Etzioni has recruited leading AI researchers from all over the world and has grown AI2 into an important part of the Seattle ecosystem. Recently they formed an incubation arm for launching new companies, and Madrona has already invested in two of their companies. In addition to founding AI2, Allen, before he died in 2019, also founded two other important research foundations, each with $100 million endowments—the Allen Institute for Brain Science and the Allen Institute for Immunology.

Although angel investors and incubators are important to a healthy startup ecosystem, eventually a young business will need the resources of a venture firm that can invest the millions of dollars needed to grow a company from its initial idea into a business employing hundreds of people and generating hundreds of millions of dollars in revenue. The amount of venture capital eclipses angel investing, with total venture investing in 2019 of over $130 billion in about eight thousand startups.[9] Still, many businesses that went on to employ hundreds of people would never have gotten their start without angels.

Several of the most successful U.S. companies founded during the past half-century—including Seattle's Starbucks and Costco—never had venture funding. Some founders make much of this fact, proud to announce that they never had to obtain venture funding and recommend against it.[10] A reason for eschewing venture funding is sometimes a belief that venture investors often have shorter timelines for realizing on an investment and will push to go public or sell to a larger company. Indeed, venture investors will someday want liquidity, but this has often stretched in recent years to ten years or more. In my experience, the CEO/founder usually has the deciding voice in how and when to achieve liquidity. Obtaining a venture investor can mean giving up a larger share of equity and potentially losing control. But venture-free companies are rare. Almost all large tech firms are funded at some stage by venture capital because building a robust company requires the kind of management and strategic assistance a venture fund offers, as well as more capital than angels alone can provide.

Take Amazon. Although initially funded by angels, a second, much larger round of funding came from venture capital. In the spring of 1996, less than a year after Amazon had launched its website and closed its angel financing, I came home from work to hear my wife Judi asking, "Do you know some guy named John Doerr? He's been calling every fifteen minutes, and he wants to invest in Amazon." Doerr, the managing director of Kleiner Perkins, was arguably the leading Silicon Valley venture capitalist at that time. He had led Kleiner's initial investment in Netscape in 1994, and as a board member helped the internet browser company go public in 1995.[11]

Kleiner invested $8 million in Amazon in 1996 at a $40 million valuation.[12] This was the only venture round for Amazon. In the spring of 1997, when Amazon went public, it raised $54 million at a valuation of $438 million. With an IPO price of $18 per share (not adjusted for subsequent splits), the value of the $1 million of angel investments was $54 million and

Kleiner's $8 million investment was worth $61 million.[13] Very nice returns, but they grew much bigger after the IPO.

Although Madrona initially was a group of angels, we gradually acquired experience as investors, and we now look back on ourselves as "professional angels." Greg Gottesman was a big help in scouting for new companies, performing due diligence investigations, reviewing their business plans, products, and anticipated customers. Soon our investment opportunities began to outstrip our resources.

After our first four years, I had a conversation with Jay Crandall, head of Bob Bass's large private equity firm on Sand Hill Road, with whom we had a working relationship. Crandall politely told me that Madrona didn't seem to be very successful at private equity investing (an understatement) but seemed pretty good at venture. He suggested that we concentrate on venture. If we raised a venture fund with outside investors, he said Bob Bass would invest several million dollars.

My three partners and I liked being private investors, but we were having difficulty keeping up with the opportunities and believed we needed to grow our team. We were having enough investment success to think we could raise at least a small fund. So in the summer of 1999, we created a formal venture capital limited partnership to raise money from outside investors. Our goal was to raise a fund of $100 million. But these were hot times for investing in tech startups, and we had a growing reputation for being able to identify successful ones, resulting in a good investment track record. Very quickly, by December, our initial venture capital fund grew to $250 million. It included Bezos, McCaw, and Stanton; other Seattleites who'd made money from tech; several families who controlled local industrial companies; and a few institutional investors like Morgan Stanley, the Hillman Company, and Microsoft. We received investments from over one hundred individuals, with commitments ranging from $100,000 to $5 million. Our original four invested the largest share.

We immediately began investing, but within months the economy experienced what was called the tech bust, as the NASDAQ peaked in March 2000 and globally we entered a severe recession in 2000. Many of our companies performed poorly. We performed triage, explaining to at least half of our companies that we would not continue to fund them. Based on our conservative accounting valuations, the value of our fund initially declined even though we remained confident that our best companies would carry us to a strong return in time. Three of our investors asked for their money back but relented after my tense conversations with them. Our investors continued to meet our capital calls, and we were able to continue to fund our best companies and make new investments. Over the next eight years, the fund was able to distribute to investors almost two dollars for every one dollar of investment, putting us in the top 10 percent of all U.S. venture funds for the 1999 vintage.

In 2003, with paper losses still showing, we needed to raise a new fund. Some investors were not impressed with our performance, and many had soured on investing in venture capital funds. Our future as a venture fund was in jeopardy. My fellow managing directors and I flew around the country presenting to potential institutional investors. I remember flying into Charlotte one night in a lightning storm to meet with a firm that managed funds for a group of small New England colleges, wondering whether this was worth our effort. Fortunately, they invested. I met with my alma mater, Harvard, which did not.[14] Neither did Yale. In the end, the strong personal and business relationships I had with McCaw, Bezos, and Doerr helped make a difference when they served as references. For example, when Chris Douvos at TIFF (The Investment Fund for Foundations) pestered Doerr for his opinion of Madrona, he responded by email: "KP does not allow me to invest in other venture firms but if I could it would be with Tom's firm. IMHO they are the best in the Northwest."[15] TIFF invested.

A similar positive recommendation from Bezos helped with another institutional investor: "Tom was an early investor in Amazon.com and consistently gives us not only 'advice' but 'help' which, of course, is always more, well, helpful. . . . Tom saw something in the idea of Amazon long before most."[16]

That said, relationships were not always enough. I called on Steve Ballmer, who was by then CEO of Microsoft, to ask them to reinvest. In vintage Ballmer, he raised his voice (a mild response for him) and bellowed, "Why would we invest in you just so you can steal our best employees?" I pointed out that we create companies that Microsoft is sometimes happy to buy, but he was not willing to be convinced.

In the end, we raised $169 million, significantly less than our prior $250 million fund, but it allowed us to stay in business. In the seventeen years since, we have gone on to raise over $2 billion, lately at the $350 million level, in eight venture funds, which we have invested in more than two hundred startups, primarily in the Seattle area. More recently we also launched acceleration funds for later-stage startups and have raised several hundred million dollars for them.

Not every investment has been a home run—and they don't all have to be. You learn something from each investment, and you apply it to the next one. We've also had significant successes, including Nordstrom.com, Isilon Systems (a developer of hardware and software systems for storing massive amounts of digital content), ShareBuilder, Smartsheet, Turi, and Snowflake.

A successful venture firm depends on building and renewing the quality of its team, belief in the power of innovation to change the world, willingness to take risks, and single-minded commitment to building great companies. To accomplish this means that we have to continually recruit a dedicated group of managing directors, other investment professionals, and support staff who work every day to identify new investments and help our companies grow. Our continued success is very much due to their quality. A firm that does not continually renew itself by hiring and promoting talented younger people is, in my view, headed for its decline.

In 2000, Paul Goodrich convinced Matt McIlwain to move with his family from Atlanta to Seattle to join Madrona. A natural leader, McIlwain has been essential to our growth. Over the years, we also convinced stellar individuals to join us: Tim Porter (Teledesic, Microsoft), Scott Jacobsen (Amazon), Len Jordan (Real Networks, Frazier Venture), Soma Somaseger (Microsoft), Steve Singh (Concur), and Hope Cochran (Clearwire, King Digital).

Together, our team powers Madrona's flywheel that we believe has helped make us successful (figure 6.1).

A key element in the venture flywheel for building long-term value is often overlooked by outsiders. Selecting promising companies is important, but even more important for most investments is helping the company build value through strategic advice, identifying and recruiting key hires, introducing potential customers, structuring and negotiating follow-on private investment rounds, and helping on exits through IPOs or corporate sales. Discussing the many nuances of this assistance would be a book in itself.

Before Madrona, Seattle had several smaller venture firms that were important to building Seattle's tech ecosystem, even though they did not survive past the 1990s. Cable & Howse, created in 1977, invested approximately $160 million in one hundred companies, providing essential early funding for Immunex, the first important biotech firm in Seattle. OVP, originally Rainier Venture Fund, was created in 1983, partially funded by a local bank, Rainier National Bank, and Hambrecht & Quist from San Francisco. By the time it wound down in 2012, OVP had invested over $750 million in 130 companies.

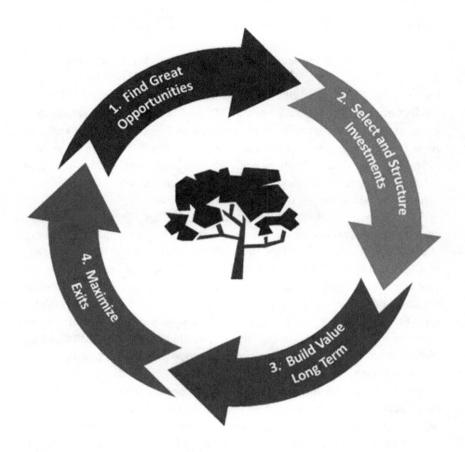

Figure 6.1
Madrona Venture flywheel *Source*: provided by the author.

OVP benefited from unusual investments of over $100 million by the State of Washington Investment Fund, which normally does not support venture funds without a lengthy track record.[17] Other states, however, do allow their pension funds to consider economic development in making investments. Alongside a more traditional portfolio, they will invest in venture funds focused on local startups. Oregon has used this approach effectively. States that want to encourage funding for startups should consider doing likewise.

Today Seattle has a healthy mix of angels, professional angels, and venture firms, although a common belief among local commentators and technologists is that Seattle would benefit from having more Madrona-sized venture funds.[18]

Soon after Madrona launched its 1999 venture fund, two others were founded: Maveron, which leverages the connections and reputation of Star-bucks' CEO Howard Schultz to invest in a wide array of consumer and retail companies across the United States; and Ignition Partners, launched by several former Microsoft executives, which invests in Seattle and Bay Area IT companies. Both have raised a series of $200 million-or-greater funds. Sometimes they compete with Madrona for deals, but we also have made many coinvestments with both. Seattle also has several smaller venture funds, each with less than $100 million to invest.

Other than bad luck or poor timing, there are several reasons that venture firms do not survive. The most substantive are a failure to recruit and retain high-quality finance professionals and a lack of sustained success in investments. A major factor behind Madrona's success has been our recruitment and promotion of smart, hard-working associates, as well as experienced tech executives, some of whom have become managing directors. Others serve as venture partners and strategic directors. In recent years we have also recruited a group of specialized professionals in hiring, public relations, and business development to assist our companies.

A serious allegation against the venture capital industry, and all of corporate finance, is that for too long it has been a white male club. The only exception is that Asians, and Indians in particular, have risen to the top of the venture and startup worlds. At least 20 percent of the founders of Madrona's companies are first- or second-generation immigrants. In the last few years, we have begun to make incremental progress bringing women into venture firms, startup leadership roles, and boards of directors. We have seen that women provide valuable new insights and have made us better investors and mentors. Venture investing in women-led startups is increasing, but the persistent absence of Blacks and Hispanics is even more striking. According to a 2018 report, 4 percent of the venture industry is Black compared to 13 percent of the U.S. population. Fully 81 percent of VC firms don't have a single Black person in an investment role, and fewer than 2 percent of partners are Black. Black founders receive less than 1 percent of all VC funding.[19]

Venture funds, with their skills at developing entrepreneurial talent and deploying capital in support of unproven businesses, are in a key position to advance the economic opportunities for women and people of color. This will require years of effort, and we are already late in starting.

As a first step at Madrona, we launched a program to educate all of our team members on unconscious racism, using an excellent online program that Microsoft developed and made available free of charge.[20] In recent years, we have increased our hiring of women and now are making a like effort to hire people of color, but we are only at the beginning.

Second, with our over eighty portfolio companies, we are encouraging and assisting them in developing their programs to increase opportunities for women and people of color.[21] Madrona and other venture firms are at an important intersection of matching companies who are searching for new employees and candidates seeking jobs. At Madrona, we spend a great deal of time helping our companies identify prospective hires, recruiting new employees, and mentoring leadership teams. We have three full-time professionals helping our companies with recruiting. We are constantly updating our database of jobs and prospects, which we are expanding to include more women and people of color.

Third, we and other venture firms need to increase our mentoring and support of minority entrepreneurs, leaders, investors, and board members. One initiative that we actively support, Onboarding Women, is serving as a model for a program to increase racial diversity in leadership positions and on boards. Onboarding Women has been successful in increasing the percentage of women on public and private company boards in our region. This program includes training and lectures by leading entrepreneurs and executives, networking, building a database of potential women board members, and making connections with companies looking for board members. In early 2021, the Black Boardroom Initiative was launched in Seattle to identify individuals to participate in six-month programs, providing them with substantive, targeted training and connections.[22]

We also have encouraged our companies to join in the national Director Challenge pledging to add a Black member to their board of directors.[23] So far, three of our companies have followed through by adding Black board members.

Fourth, the venture community must expand the availability of capital for women- and minority-led companies. There are several women-led venture funds in the United States, and a local company founder and angel investor, Jonathan Sposato, makes it a policy only to invest in women-led companies. Sposato has written an excellent and relevant book—*Better Together: 8 Ways Working with Women Leads to Extraordinary Products and Profits* (2017).

In addition to each venture firm making a greater effort to find and invest in women- and minority-led companies, we also need to consider whether a new investment vehicle in our region would help give minority investors the necessary experience and increasing minority investments. This could tie into efforts by groups like Pivotal Ventures (a Melinda Gates–backed fund) and Black Future Co-Op Fund. There are several ways to set up this type of venture, but one example to consider is Andreessen Horowitz's Talent x Opportunity Fund, structured as a donor advisory fund that will reinvest all profits in financing future entrepreneurs. It focuses on entrepreneurs from underserved communities "who did not have access to the fast track in life but who have great potential" and provides seed capital, training, and network connections.[24] Madrona is supporting Microsoft's Startup Success 2020 program, which offers content, connections, coaching, and community for twenty-seven tech startups led by people of color.

But all of these programs will not be sufficient if we don't make it possible that many more people of color qualify for technology jobs by improving education for minorities at all levels—preschools, K–12, and universities. This is why we and others are supporting organizations such as Rainier Scholars, Technology Access Foundation, and Washington STEM, which are building the pipeline for future minority scientists and entrepreneurs.

As discussed earlier, the first step for a city that wants to develop a tech economy is to increase the number of angel investors willing to invest in local startups. More difficult is the creation of local venture firms. There are several possible routes to take, but a usual starting point is a small group of individuals with experience in business or investing who focus on startups. After realizing enough investment success, they transition from angel to venture capital by attracting commitments from outside investors. This is precisely what we did at Madrona. We worked for five years together before raising our first venture fund. The challenge is that this early phase typically takes several years, and without sufficient funds, it can be difficult to build a track record. It's a chicken-and-egg problem: you need talented people working together for several years to realize successful investments and raise capital, but you need adequate capital to attract and retain talented people.

Increasingly, however, young venture firms are spreading beyond traditional centers of tech investment in the Bay Area, Boston, and Seattle, into a wider range of locales—a very promising development, in my view. Venture firms are also looking beyond their local city limits for investment opportunities in other regions—particularly if a faraway startup company is generating some customer interest.

Why haven't other cities similar to Seattle developed into important tech centers? Many have strong universities, global 500 headquarters, and accumulated investor wealth. Honeywell and General Mills were based in Minneapolis, McDonald Douglas Aircraft and Monsanto in St. Louis. It wasn't luck that made Seattle successful, nor the city's high quality of life—others have that too—but the foresight of a small group of tech pioneers who consciously set out to build a tech economy. It took years. But it's not too late for other cities to initiate or accelerate their efforts. Focus on building your flywheel.

Angel and venture investing have persisted because this type of risk capital has provided the essential funding for many successful tech companies, and the overall investment returns over time have been better than returns from most asset classes. This is not to say that every angel investor or venture firm makes lots of money. An angel could easily make five to ten investments and suffer a loss overall. Likewise, about half of all venture funds lose money, and only a few achieve the standard industry target of 20 percent IRR or three times the return of capital.

Yet the best venture firms often continue in achieving superior results, so there is more to the pursuit than luck. In truth, success often breeds success, since a successful fund will get the best opportunities to recruit and retain talented investment professionals who are likely to identify the most promising companies. Still, even the best firms may lose money on a third of their investments. But because the maximum loss on an investment is limited to the capital invested, and the maximum return can be as much as five or even ten times or more, a fund that experiences many losing investments can still achieve positive returns overall.

Today, when people ask what I find most difficult about investing, I tell them my challenge is reining in my enthusiasm for new ideas. For an early-stage tech investor, I consider this more of a virtue than a vice.

7

New Models for Success
Oklahoma City, Tulsa, and Kansas City

A recent Seattle news story was headlined "Seattleites Don't Realize It Yet, but It's Pronounced Tech-Oma," referring to the nearby city of Tacoma of two hundred thousand people. The story highlighted the growing interest of Seattle tech startups moving to Tacoma and rumors that Amazon was looking there for additional software office space.[1]

In February 2020, just before COVID-19 hit, two entrepreneurs from Tacoma visited me at Madrona to pitch on investing in their local startups because their city was beginning to attract more tech workers and generate promising startups. It was mostly an uphill argument because even though located only thirty-four miles south of Seattle along the I-5 corridor, Tacoma was not even a minor tech center. But now, one year later, the office landscape is changing.

In the early days of the internet, some of us imagined that workers could work from anywhere and that this would stimulate tech workers and companies to locate in smaller cities near Seattle. Bill Gates Sr. and I, representing the Technology Alliance, had a slideshow we presented to chambers of commerce in secondary cities on the tech opportunity. But there was little movement, as tech workers wanted to congregate together in Seattle. Now with the trend for workers and companies being more interested in locating in smaller cities, and partly motivated by worsening social problems in tech cities, the Tacomas of the country have accelerated opportunities to get their flywheels spinning.

Greg Shaw[2] and I left Seattle for Oklahoma City on March 3, 2020— Super Tuesday—to see how three midsized, Midwestern cities were getting their flywheels underway. By dinnertime over fried catfish and cornbread, Joe Biden had been declared the winner of Oklahoma's primary and of the night's contests around the country. Politics, however, would soon withdraw from mainstage to small screens and Zoom calls. Over the weekend, a man living just outside Seattle became the first American known to die of coronavirus, which had already begun to spread beyond the Puget Sound region. In retrospect, it's amazing we made our trip to three Midwestern towns, and that people kept their meetings with us. But they did, graciously.

During my flight, I reviewed the impressive list of entrepreneurs, startup founders, investors, accelerators, philanthropists, and community leaders I would talk with who were at work building innovation flywheels in Oklahoma City, Tulsa, and Kansas City. These may not seem like obvious candidates to learn about flywheels of tech and innovation, but I could not ignore what my research was revealing. Like any coastal elite, I had to confront my own biases.

I view with some suspicion most rankings of "best cities." Some measure subjective lifestyle (Honolulu) based on favorable Facebook mentions, or they depend too much on data like the number of patents received. I wanted to see for myself what was happening. Tier-one regions like Silicon Valley and Seattle produce innovation on a steady basis. And Austin and Pittsburgh are also producing results. But something is beginning to stir in many other cities.

To build a tech flywheel, a good place to start is the local university. Does it have a computer science department or medical school? Most have professors who are sources of important research, and, if they are freed from traditional restrictions inhibiting entrepreneurialism, they can be encouraged to turn their innovations into startups rather than licensing them to big companies.

Madrona Venture has funded twenty small companies that came out of the Paul G. Allen School of Computer Science & Engineering at the University of Washington, and biotech venture capitalists and angel investors are funding a half-dozen startups from professor David Baker's protein lab at the UW School of Medicine. Madrona cultivates its relations with leading professors and graduate students, helping them think through whether they have research that could launch a company and then helping them raise equity and build a team. Local entrepreneurs and even traditional businesspeople can do likewise at their hometown schools.

Civic leaders have roles, too. They can lead efforts to convince legislatures to loosen archaic conflict-of-interest and sabbatical limitations on professors who want to start companies, as we did in Washington State. They can also lead local fundraising campaigns for building new science facilities, hiring talented faculty, and attracting graduate students with scholarships and research grants. Based on a single email from Ed Lazowska, chair of the UW Department of Engineering and Computer Science, Bezos agreed to donate $2 million to endow two chairs so that the UW could successfully lure two professors from Carnegie Mellon. The UW is savvy about marketing its openness to faculty startups in competing with other universities for entrepreneurial professors. But funding for such positions needn't come solely from technology CEOs; any local businessperson from any industry could do the same. In Washington, for instance, the UW has been supported by traditional companies like Costco, PACCAR, and Weyerhaeuser, in addition to Microsoft and Amazon.

A town with a college, even if it is not a major research institution, can attract people who want the intellectual and social benefits of these communities. Bozeman, Montana, with the campus of Montana State University, is a good example of what attracts young software engineers and others who can work remotely. Many talented people move there to raise their families without the livability problems now plaguing Seattle and the Bay Area. Burlington, Vermont, ranked fourth on a list of the ten small cities with the biggest surge in remote workers since the start of COVID-19. It didn't surprise me to note that Tulsa was also on the list.[3]

Scrappiness, grit, and plain old-fashion hard work are the bedrock of the places we went to visit. But it takes more than just these qualities to produce a tech-economy flywheel. A wide variety of digital and next-generation skills are necessary. People with these talents are attracted by restored downtowns with a diversity of people, restaurants, and shops—a vibrant culture. They want good schools for their children and children-to-be. One trait I noted almost immediately is that these cities are welcoming back boomerangs, natives who'd left to gain education or experience and came home with new ideas and skills. They often have far-sighted political leaders. It also helps if the wealthy with capital from an earlier flywheel are willing to invest in the next flywheel. These cities, I suspect, are willing to experiment and then prioritize what works.

My first meeting was an insightful breakfast with two movers and shakers in the Oklahoma City orbit. Bob Ross runs the Inasmuch Foundation, created by the legendary Gaylord family, owners of *The Oklahoman* newspaper and Spike TV (formerly the Nashville Network). Renzi Stone is an entrepreneur and, until recently, a member of the University of Oklahoma's board of trustees. You can't miss this 6-foot-9-inch former OU Sooners basketball standout. Not only is his height notable from a distance in this prairie city, but so is his focus and energy.

The son of an entrepreneur, Stone runs a successful communication and marketing firm and devotes substantial time to build Oklahoma City's entrepreneurial infrastructure. He is working to attract the blend of seed and venture capital, incubators and accelerators necessary to complement the slowing economic engine once powered by oil and gas extraction. To that end, Stone is building a social club of sorts that will host offices and gathering places where entrepreneurs and social innovators can mix.

Their downtown has come alive again as a thriving entertainment sector but with blissfully light traffic and space for businesses to grow. The night before I met with Ross and Stone, the downtown was packed with NBA fans on their way to restaurants and to see their beloved Thunder play the visiting Los Angeles Clippers. Millennials, many of whom live downtown or in up-and-coming communities like the Plaza District, Skid Row, or Midtown, walked back to their condos after a workout at fitness centers or the river trails. Nearby, the hip neighborhood of bars and restaurants known as Bricktown has risen from the dusty, rundown warehouse district across the railroad tracks.

The urban renewal of Oklahoma City is well documented in then-mayor Mick Cornett's book, *The Next American City*. Mayor Cornett, who met me at a Starbucks across the street from the classy AAA baseball stadium (an affiliate of the Dodgers), writes that the story of renewal is less "Good to Great" and more "Worst to First."[4] Looking back, Cornett recalled he had learned that while Oklahoma City was in the midst of an effort to lure United Airlines to locate a job-rich facility, the commercial airline sent midlevel execs to spend a night there. The executives returned to Chicago and promptly chose Indianapolis as the site for the new facility. They could not imagine living in Oklahoma City, much less spending another night there.

Inspired by Denver mayor John Hickenlooper and Chicago mayor Richard M. Daley, Cornett worked to change Oklahoma City. Over his three terms as mayor, Oklahoma City invested $5 billion of public and private money, creating what *National Geographic* called the "pride of the plains" on its 2015 list of the best vacation destinations in the world.[5] Cornett credits the citizens for rallying behind the Metropolitan Area Projects Plan (MAPS), approving a sales tax that helped to fund Bricktown, school renovations, a new central library, and a new performing arts center among other projects. He paused to look out the window and pointed to the Plow Building and other structures that now house a range of startups and enterprises. He humblebragged that 150 city delegations have come to Oklahoma City to learn how they did it. "They see us as attainable."

"Place matters," Stone told me that morning. "On the other hand, it doesn't matter at all. People have the choice of where to live. Talent will find its way here if we offer the right incentives."

We also meet with Piyush Patel, who agreed. In 2000 the Oklahoma native founded Digital-Tutors, which trained developers to make movies and videogames for platforms like the Xbox. Patel sold the company in 2014 to Pluralsight for a reported $45 million. He is now an angel investor, active in business groups like the Entrepreneurs' Organization, and, like me, a winemaker. Patel summarized his view of the local benefits.

"What's unique about being in Oklahoma is you've got this amazing cost of living. We can bring in talent from LA, from Florida, from New York, outside of the country, incredible tech geniuses, and they can live a really nice lifestyle."

He told us the story of an employee he recruited to Oklahoma City from Sony Pictures in California. In LA, this recruit had complained, he routinely left home each day in the dark of early morning and returned home, also in the dark, because the commute took so long—often ninety minutes, though he was only twelve miles away from the office.

"Look," Patel told him, "you come to work here, you're going to start at nine in the morning so you can drop the kids off, and you will be home by 5:30." After his first day, he called Patel at 5:45 to say, "I'm just walking in the door. This is amazing."

"I'll be honest with you," Patel continued. "It's really easy to become a really big fish in this tiny little pond. If we were in LA or San Francisco, I don't think we would've made it. I don't think we would've been able to hold our employees or retain them because they would have had so many other opportunities. And I don't think we would've grown as fast because we would've constantly been chasing employees—'We've got to have more people, how do we keep our people?' versus, 'How do we grow our business?'"

Stone, Ross, and Patel all agree on one area where Oklahoma City needs to invest and push harder: improving educational opportunities. Ross pointed to a teachers' strike in 2018 that made national news and coincided, not incidentally, with a continued flight to the nearby suburbs of Edmond, Yukon, and Moore. "Entrepreneurial talent is outpacing our educational capacity," Stone said.

Patel would like to see tech training organizations like General Assembly and Galvanize coding boot camps in Oklahoma City. A step in the right direction is the Francis Tuttle Technology Center, one of Oklahoma's twenty-nine technology center districts providing tech training for high school students and adult learners.[6]

Oklahoma City has a strong base of engineers from their oil and gas extraction industry and an aerospace industry growing out of the Tinker Air Force Base located in Oklahoma City. Agricultural sciences can also contribute to their flywheel.

We next sat down in a grand old hotel with Patrick Fitzgerald, a native Oklahoman who left to start his careers at Campbell Soup and Citicorp. Along the way, he helped Steve Jobs and Apple build iTunes and went on to Disney to participate in their growing streaming business. When he returned home to help the city and state he loves, the lieutenant governor—an elected position charged with economic development—asked Fitzgerald to study Oklahoma's success at producing entrepreneurs. Too often, economic developers are focused on trying to attract large enterprises to open an office, a warehouse, or a data center. "Don't go for the Redwoods," Fitzgerald advised. "Get some saplings and nurture them."

Fitzgerald turned to the Kauffman Index,[7] published by the Kauffman Foundation, which ranks cities on a variety of entrepreneurial factors. The Kauffman Index showed Oklahoma City ranked high for startups but low for moving them beyond to a growth phase. On Friday we visited the Kauffman Foundation in Kansas City to learn more about their efforts to support entrepreneurs.

More would be needed, Fitzgerald realized. More incubators, more enlightened capital to help fund businesses moving from startup to businesses at-scale. There is no shortage of wealth in Oklahoma, thanks to oil and gas. Crude was discovered in 1859, and by statehood in 1907, Oklahoma produced the most oil of any state or territory in the United States, giving rise to giants like Conoco, Phillips, and more recently Chesapeake Energy and Devon Energy.

We went to talk with Walt Duncan, who leads Duncan Oil Properties, founded in the heydays of the 1950s. He greeted us in a formal office lobby featuring stately oil paintings his grandfather purchased in better times. Inside his office, geological surveys of the Permian and Anadarko basins, some of the most productive oil reserves in the world, adorned the walls. Duncan has invested in technology to help with precision exploration, but declines in oil and gas prices are limiting his ability to invest.

No stranger to boom-and-bust economics, the state was experiencing a collapse of energy prices when we visited. A month later, the *New York Times* would report that "companies are laying off workers, shutting down wells and preparing for a prolonged slump as oil prices tumble."[8]

Hydraulic fracking and horizontal drilling remain important parts of the local economy, but a new generation of wildcatting entrepreneurs is needed.

My next stop was a startup accelerator near downtown called Stitch-Crew, led by Erika and Chris Lukas. They are located in the hip Thunder Launchpad underwritten by the NBA's Oklahoma City Thunder. We joined a lunchtime gathering of fifteen angel investors interested in the local startup scene.

While investors and prospective investors enjoyed sack lunches, Erika reviewed the various venture capital funds in the area—all relatively small. These included Cortado Ventures, the Oklahoma Life Science Fund, Square Deal Capital, and others. In a few weeks, they were expecting America Online founder Steve Case's Rise of the Rest Fund to visit to hear pitches and, hopefully, invest capital in a local startup or two.

During Q&A, Erika admitted there was a shortage of capital for new tech companies. So StitchCrew also works with crowdfunders like Kiva, nondilutive capital like traditional loans, and equity buybacks.

"We've had to do a lot more with less," she said. "We have gritty founders."

That grit is prominent in the story of the legendary Sooners, the Oklahoma pioneers who jumped the gun in a sanctioned land run to settle this state.

We're reminded at our last stop of just how gritty by the bombing of the Alfred P. Murray Federal Building on a spring day in April 1995. Ross guided us on a tour of the memorial, a museum devoted to remembering "those who were killed, those who survived, and those changed forever."[9] Outside the museum, we were greeted by the Survivor Tree, a sign of life amidst such terror. Recalling our conversation with Patrick Fitzgerald,

I noted that the tree was a mere sapling in 1995 but today is a big beautiful American Elm.

~

We turned our rental car north and headed for Tulsa, another oil town located about two hours' drive through tribal lands along a neatly kempt turnpike. I looked over the stands of hardwood trees and the occasional creek, reflecting on the day. Mayor Cornett described the Oklahoma City flywheel as aviation, oil and gas extraction, tourism, a network of north–south and east–west highways that cross the country, and government employment. More than one hundred thousand college students are within a one-hour drive.

In many cities throughout the country, we are seeing an amazing amount of urban restoration and renewal—downtowns full of restored buildings from the industrial age, public parks, and attractive new structures—which are followed by restaurants, bars, and low-cost, chic apartments that in turn bring in young people. These elements and more are helping start a flywheel in Oklahoma and elsewhere.

Ahead, rising from the plains, we saw the skyline of Tulsa, and I was reminded of something Mayor Cornett said: "Do everything you can to support people who lead change because change is hard."

~

Few understand this better than George Kaiser, an energy and banking billionaire and philanthropist in Tulsa who is as at ease with Bill Gates, Warren Buffett, and Nobel Laureate Bob Dylan as he is with preschoolers, troubled youth, and prisoners. Kaiser's Jewish family fled Nazi Germany in the 1930s and settled in Tulsa, an unlikely choice given the city's fundamentalist Christian bent (it's home to Oral Roberts University), troubled race history (site of America's deadliest race riot), and heritage as home to cowboys, Indians, and roughnecks. But Kaiser and his capable team at the Kaiser Family Foundation are well on their way to changing that image to a city some regard as the next Austin or Nashville. If Austin and Nashville are synonymous with country music, Tulsa should be equated with Americana. It has inspired many a singer-songwriter—"Tulsa Time" and "Take Me Back to Tulsa," to name a few. It is home to the Woody Guthrie archive and museum, and soon Bob Dylan's archive.

As I sat down to dinner with Kaiser's team, the pop band Hanson ("MMMBop") stopped by our table for a chat after a session at their next-door studio. The trio hails from Tulsa and is working to build on the city's historical links with musicians and the music industry. The restaurant is located in Tulsa's historic Arts District, which the Kaiser foundation and other local partners are revitalizing as a cool, urban link between the Black community to the north and the wealthier white neighborhoods south of downtown. With its collection of artisan cafes, breweries, ethnic food, bars, and hip workspaces, anyone from Berkeley to Brooklyn would be comfortable here.

We engaged in energetic conversations over dinner and wine with Kaiser's team, led by dynamic Ken Levit and includes Robert Thomas, chief investment officer, who grew up in Washington State and worked for BGI, Bill Gates's in-house money and investment managers in the Seattle area. Thomas says he prefers living and working in Tulsa.

He is not the only one. Tulsa made global headlines when the Kaiser foundation launched Tulsa Remote in 2018.[10] Recognizing the growth of remote workers—two years before COVID-19—the foundation began offering several hundred people who can work from anywhere $10,000 and the chance to help build an entrepreneurial community in Tulsa.

The person they found to run the program, Aaron Bolzle, is a living advertisement for the innovative approach. A graduate of Booker T. Washington High School in Tulsa, he left for the Berklee College of Music in Boston and then went on to Apple iTunes in California, where he met our other boomerang friend from Oklahoma City, Patrick Fitzgerald. After a few years of fighting traffic and skyrocketing living expenses, Bolzle began to look around. He could work from anywhere, and so he investigated the usual places—Denver and Austin to name a few.

"I didn't want to be just another tech bro," he said. Instead, he moved home to Tulsa.

He had changed during his time on the East and West Coasts, and his concerns about moving back were similar to those of others who apply for Tulsa Remote: "Will I fit in? Will I meet people?" The answer was a triumphant, yes.

Tulsa Remote was soon oversubscribed with ten thousand applications from 155 countries. They must first come to Tulsa for a recruitment weekend. In the first year, 112 moved to Tulsa and have put down roots. There will be 250 in the program in the next tranche, and 2,500 are expected over the next ten years.[11] Bolzle can see several likely startups being generated from this group. He feels he is "stocking the pond" for innovative businesses.

A key feature of the program is that the foundation organizes a series of seminars and social events so that after each annual class arrives, it gets to know other alumni and build long-term relationships. Other cities should do likewise with remote workers moving to their cities. And now, $10,000 incentives are not necessary.

I met Bolzle in a comfortable coworking space called 36 Degrees North, Tulsa's self-proclaimed basecamp for entrepreneurs, located in an old Model T factory. The aim is for 36 Degrees North to evolve into an integrated entrepreneurial hub like Chicago's 1871 and Birmingham, Alabama's, Innovation Depot.

Bolzle introduced me to Taylor, a developer who helped found YouNeedABudget.com, a financial planning service. Taylor, his wife, and their toddler moved from Austin. Bearded and wearing a hip Western shirt, he told me he first learned of the Tulsa Remote program in a Slack post. He mentioned it casually to his wife one evening, and to his surprise, she said, "We have to do that." They wanted out of Austin because they felt it was becoming overcrowded.

Mary Jackson, her arms adorned with a multitude of stylish tattoos, moved to Tulsa after stints in South Carolina and Washington, DC. Friends sent the Tulsa Remote announcement to her because they knew she would embrace the idea.

"We moved here to meet the other weird people," she said with a smile.

In Okie terms, Taylor and Mary are modern pioneers, the new Oklahoma Sooners. However, as more and more young workers looked to Small Town, USA, to work remotely during the COVID-19 crisis, a national news article headlined in June 2020: "For Newly Remote Workers, Small Town U.S.A. Will Lose Its Allure Soon Enough."[12] Tulsa argues otherwise with over 90 percent of the participants in Tulsa Remote staying long term. And Tulsa Remote now allows the $10,000 to be applied to a down payment on a house. The median house price in Tulsa is $157,000.[13]

A few streets over, I visited Holberton School, founded by two Silicon Valley developers—Sylvain Kalache and Julien Barbier—to teach students with little or no computer experience to think like programmers and learn to code. Libby Wuller, another boomerang Oklahoman who left to work in Washington, DC, proudly showed us her urban office in the late 1800s brick building that once served as a warehouse. Aspiring coders in hoodies and T-shirts sat in open spaces that looked like a set for the TV show Silicon Valley. Holberton has similar campuses in San Francisco and New Haven.

About 70 percent of the students are local, and they receive a monthly stipend to help ends meet.[14] The program lasts for about eighteen months. Students learn a variety of marketable languages and disciplines. They are exposed to C, Python, object-oriented programs, dev-ops, MySQL, GitHub, augmented reality, machine learning, blockchain, app development, and cybersecurity.

The only requirements are that students be eighteen years old, have a high school diploma or equivalent, and have an aptitude to learn in the way Holberton teaches, which is assessed with a simple test. They sign an income-share agreement to pay back tuition. The terms are 17 percent of a graduates' gross income for 42 months after the student is employed in a job that pays at least $40,000 per year, capped at $85,000, or 10 percent if you remain in Tulsa. One local employer told Libby Wuller that the company could hire every graduate.

Across town, the University of Tulsa sees the same opportunity. We sat down with Janet Levitt, the president, and John Hale, head of TU's computer science program. The venerable university, long known for its outstanding petroleum engineering program, has shifted to focus on broader STEM programs. Computer science is now the most popular degree. Part of the reason for this is an elite cybersecurity program to support the National Security Administration (NSA) and other government security agencies. A big opportunity for TU and its students is if they can expand their cybersecurity program beyond the government to prepare their graduates to meet the broader security needs of businesses.

One activity at the university caught my ear—the popular gas and convenience store in the South, Love's Country Stores, sponsors the Entrepreneur's Cup, which brings together TU's business and computer science schools, to pitch new ideas for startup tech businesses. This sounded promising.

The combination of TU's cybersecurity program, the remote workers initiative, and Holberton School can all be important in accelerating Tulsa's flywheel.

But the city still has some challenges to overcome—notably, vestiges of its ugly racial history and high poverty rates in traditionally Black north Tulsa. We drove through north Tulsa (Greenwood District), a sprawling area of single-family homes mostly occupied by Blacks and small businesses, which was one of the targets of the Tulsa Race Riots of 1921, also known as the Black Wall Street (or Greenwood) Massacre.[15] Hundreds (mostly Blacks) were killed or injured. Black Wall Street was a prosperous Black business district where 191 businesses were destroyed. According to the Red Cross, 1,256 residences were burned and a further 215 looted.[16]

When we visited, north Tulsa appeared to be on the upswing—with visible home improvements and a new 72 acre industrial park that could expand into 292 acres.[17] We saw lots of large family homes for sale and can confirm in Redfin that prices are moving up.

But north Tulsa still has high poverty and unemployment rates compared to the rest of the city.[18] New jobs, improved education, and retraining will be essential. A new Amazon fulfillment center, which will employ 1,500 people with a minimum wage of at least $15 per hour plus benefits, will help. Amazon's Career Choice program for all fulfillment workers pays 95 percent of tuition for courses in high-demand fields such as nursing and IT services, so they can move up at Amazon or qualify for good jobs elsewhere.[19]

Within a few weeks of my visit, Elon Musk's Tesla automobile company announced that Tulsa and Austin were finalists for a new factory (Austin was picked, but Oklahoma governor Kevin Stitt responded by saying that Tulsa was ready for Tesla's next expansion[20]).

A special stop during my visit was to one of four early learning centers funded by the Kaiser Foundation to provide year-round, high-quality care and education programming to over five hundred two- to four-year-olds in low-income neighborhoods. Early learning is a major focus of George Kaiser and his foundation. He believes that early childhood education is a critical step in improving educational outcomes for children, particularly from low-income families.

I also visited the small Greenwood Cultural Center, "the keeper of the flame for the Black Wall Street era,"[21] whose energetic director explained their plans and dreams for expansion, hoping that I would put in a good word with the Gates and Bezos foundations when I returned to Seattle.

My day in Tulsa ended with a magical late-afternoon walk with George Kaiser through the Gathering Place, a world-class one-hundred-acre park funded by his foundation and quickly named by *Time* magazine as one of the world's greatest places. The park attracted an average of 2.5 million people each year during its first two years.[22]

Kaiser, Tulsa's great benefactor, understands his city's fraught history and its challenges. As we strolled through the Gathering Place, he told me, "We got divided by race, ethnicity, politics, and geography. The Gathering Place is meant to help unite."

The park is set along the wide Arkansas River with views of the Tulsa skyline on one side and the city's historic oil refineries and tank farms on the other. It was early March but already the sun was hot, and the park was full of mothers and children, seniors, and students just out of class. It was a remarkably diverse group. Kaiser was kind, witty, and nattily dressed as we admired the landscaping and playgrounds and in the distance the Arkansas River, refineries, and tank farms. Kaiser wanted a distinguishing characteristic for Tulsa, something that might help to attract and retain entrepreneurs and employees. He wanted kids to have a nostalgic memory of Tulsa where they will want to bring their children.

We approached a boathouse with the soaring white sails reminiscent of the iconic Denver Airport, an enormous playground with whimsical geese designed by a Danish artist, and a waterpark that encourages kids to communicate and collaborate.

We ended up at the Williams Lodge, an inviting building with tall windows. Anyone and everyone in the community is invited to come and spend an hour or a whole day. It's a stunning space where we see students working on laptops, seniors playing cards, kids playing. In a long room overlooking the playground, Kaiser and his team had scheduled several meetings, and we were invited to join.

First on the agenda was to review designs for the new Bob Dylan Archive, which the foundation acquired for $20 million. Dylan had approached Kaiser after the opening of Tulsa's Woody Guthrie Center several years ago about creating something similar for his collection. The two educational institutions will be next door to one another in the arts district when it opens in 2021. We watched never-before-seen footage of Dylan touring and making music.

Next on the agenda, fortuitously, was a discussion with the management consulting firm McKinsey & Company, which had been working at Kaiser's request on how best to create new jobs in Tulsa. Inspired by former Seattle mayor Charlie Royer, Kaiser was curious to know what are the five qualities that cities like us need to have to be successful? What are the kinds of jobs that we should create? How do we create a next-generation economy?

We saw only a fraction of the answers to those questions. On that particular day, they focused on leveraging their healthcare assets to create jobs. Specifically, they recommended four areas within healthcare to concentrate on—telemedicine, mental health research, health informatics, and social determinants of health. How do we engage and attract a myriad of employers, build a community, and create an innovation pipeline? Though overly elaborate, the presentation provided a good roadmap for any community to follow.

Creating a flywheel and keeping it moving requires many people and institutions working for common causes. But there is also no ignoring the impact that a single person like George Kaiser can have on behalf of a hometown. Yes, a person with considerable resources but also love, passion, and know-how. A Seattle version was Paul Allen. They should both be an inspiration and a guide for leaders and hometowns everywhere.

∽

The Kansas City flywheel was not immediately clear as we landed in the city that straddles two states (Kansas and Missouri), but it became more transparent over our many conversations.

Kansas City is larger than Oklahoma City and Tulsa, but all three share an entrepreneurial culture. Jonathan Kemper, whose family founded the venerable Commerce Bank 150 years ago, invited us for dinner at a bygone-era country club with his family to discuss that entrepreneurial spirit. Kansas City is home to numerous world-class engineering and architecture firms, and to Sprint, Hallmark, H&R Block, Garmin, Russell Stover Candies, and Marion Labs (which became Marion Merrell Dow). Marion was founded by Ewing Kauffman, whose wealth produced the Kansas City Royals, and most significantly for this story, the Kauffman Foundation, whose grantmaking is focused on generating momentum for an innovation flywheel—education, entrepreneurship, and civic support.

The Kemper family built Commerce Bank in the late 1800s and watched the city produce more than its share of global businesses. Along with other pillar families, they are involved in many aspects of the city—from community development and affordable housing to economic development and entrepreneurship.

Joining us for dinner was Jonathan's wife and two of their adult children, Charlotte and David. All are engaged in the city's entrepreneurial future—David in urban renewal and affordable housing, and Charlotte in finance and support for innovative startups. Over dinner, Jonathan told the story of Almon Brown Strowger, a Kansas City itinerant who created what those in the telecommunications industry call the Strowger Stepping Switch. Having been involved in the telecom industry, my ears perked up as I listened to Jonathan tell the Strowger story. As legend has it, Strowger owned a funeral home and was losing business to a rival whose telephone operator wife listened in on party lines to learn in real-time who had just died. She relayed that information to her husband, who pounced on the new business. Frustrated, Strowger investigated a way to remove the intermediary, creating a stepping switch that removed the nosey operator and neighbors. Local lore is that this led to Sprint, one of America's innovative phone companies. I have no proof, but I do know that freewheeling can work in surprising ways. Similarly, lore has it that the engineering know-how that made it possible to bridge the mighty Missouri, which divides Kansas City, led to a plethora of locally headquartered global engineering firms.

The next morning, Charlotte joined us for an insightful breakfast meeting she had arranged with Neal Sharma and Sherry Gonzales in Kansas City's historic Plaza District, home to the world's first shopping center designed for customers arriving by car. Its 1922 architecture was inspired by Seville, Spain, seemingly a far cry from our images of shopping malls. (Curiously, a little-known correspondent for the Kansas City Star, Ernest Hemingway, would go on to make Spanish culture quite famous.)

Gonzales is an executive with the Civic Council of Greater Kansas City and runs KC Rising, an effort to accelerate the region's economic growth. Sharma chairs the group and is the founder of the Digital Experience Agency, or DEG, a top digital marketing firm.

Sharma feels that Kansas City's "Midwest ethos"—an attitude for carefully studying problems before committing to action—is an advantage. "More Warren Buffett than New York hedge fund," he said. But that Midwest ethos also can be risk-averse, investors preferring to stay within their lanes of comfort—a disadvantage for accelerating a flywheel.

In 2006, the Brookings Institution published a report that found the Kansas City region's innovation capacity was weak. It pointed out that the region's division between two states interfered with coordinated efforts and reported that the region still suffered from a legacy of severe racial inequality. The community took the report seriously and asked Brookings to return in 2014 for a follow-up. Researchers noted progress had been made in some areas but still the economy was less dynamic than it needed to be.[23]

Families like the Kemper's, the Kauffman Foundation, the Kansas City Civic Council, and now KC Rising are working together to generate momentum for their region's flywheel. They have seen some encouraging successes, including the founding of a venture capital fund, which is important not only for the funding it provides entrepreneurs but also for the collective learning it provides investors. It's a so-called sidecar fund that coinvests with institutional VCs, like my own, from around the country. They are working together to create advisory boards to help support entrepreneurs and an innovation hub. Promising tech businesses have begun to come out of this silicon prairie—from the financial tech company C2FO to PayIt, Rx Savings Solutions, and a service that I would like to use—an Uber-inspired rental of pickup trucks.

More recently, KC Rising is articulating a new focus on leveraging sector strengths, such as the region's heritage in architecture and engineering, its location in the middle of the country for logistics, and helping startups grow to scale.

If there is such a thing as a shrine to the nobility of entrepreneurship, it's located in Kansas City at the Ewing Marion Kauffman Foundation. Visitors to the foundation's campus are greeted by a plaque that communicates the Entrepreneur's Pledge:

I am an Entrepreneur. I am following a dream, pursuing an opportunity, taking charge of my own destiny. I am bringing something of value to society, making a job for myself and for others, and creating wealth that benefits my family, my community, my country, my world. I am one of a movement of millions of entrepreneurs and innovators who made America great, and who will continue to keep our economy going . . . and growing. I am what I am because many people have helped me along on this journey.

Therefore . . .

I will tell my story, sharing my successes and failures, so that others taking the entrepreneurial path can learn. I will strive to mentor an aspiring entrepreneur. I will make my voice heard by those who make policy decisions that affect me and my business. I will appreciate and celebrate my accomplishments, and the accomplishments of all my fellow entrepreneurs. I will give back to the society that helped me to be successful. I will Build a Stronger America.[24]

With its $2.6 billion endowment, the Kauffman Foundation is located on an understated, tasteful campus not far from downtown and the plaza area. A statue of the founders is out front. It devotes roughly 50 percent of its effort to education and 50 percent to economic development, attempting to balance national programs and local needs.

We met with Wendy Guillies, Kauffman's CEO, to learn more about entrepreneurship in Kansas City. Guillies rose through the ranks and is a smart, enthusiastic supporter of her community and foundation. Her description of the Kansas City flywheel is as simple as it is intriguing. It builds on much of what we heard:

- Designing and building, which relates to the engineering and architecture history
- Transaction and logistics, which summarizes the region's financial services background and its geography
- Work ethic and love for the community, which gets at what Neal Sharma earlier described as the Midwest ethos

Kauffman has been a pioneer in supporting entrepreneurship across the country, including creating the Angel Capital Association and funding fellows and apprenticeships to increase the presence of women and people of color in venture capital. The foundation is a sophisticated, long-term investor in venture capital firms, including Madrona.

It has worked to professionalize the field of entrepreneurship and encourages mayors who support entrepreneurship. The foundation funds accelerators to help startups in the four-state region of Kansas, Missouri, Nebraska, and Iowa, and supports innovative approaches to teaching entrepreneurship in high school and tech skills for adults. The former head of the education programs at the Bill & Melinda Gates Foundation, Tom Vander Ark, told me that the Kauffman Foundation's focus on supporting a high school program that focuses on market value assets like internships, setting college expectations, and projects for businesses—not just a diploma—is one of the most inventive he's seen. It's real-world learning, and it's now in twenty-three school districts.

One of my favorite ideas they've sponsored is 1 Million Cups. Based on the notion that entrepreneurs discover solutions and engage with their communities over a million cups of coffee, the foundation funds this free program designed to educate, engage, and inspire entrepreneurs around the country. It's now in more than 160 communities.[25]

As if all of that is not enough, Kauffman is perhaps best known for its index, an indicator of entrepreneurship trends in the United States, including series on new employers' information and early-stage entrepreneurship.[26]

Guillies concluded with an interesting analogy. Major league teams like the Kansas City Royals have a talent pipeline. They invest incredible resources in building and maintaining that pipeline of future big leaguers. Why not do the same for entrepreneurs? They are, after all, our greatest source of jobs and innovation.

At lunchtime we headed across town to one of Kansas City's many famous barbecue joints to meet with a recent Kauffman alum, Victor Hwang, who left the foundation to strike out on his own adventure. He is credited with transforming Kauffman's entrepreneur program from its previous one-on-one focus to broader programs building ecosystems to support entrepreneurs. Born in Baton Rouge, his two computer science parents started a consulting business to help fund his education at Harvard.

Hwang is coauthor of *The Rainforest: The Secret to Building the Next Silicon Valley* (2012),[27] a book he described as "about the nature of innovation . . . about the nature of complex innovation systems—whether in Silicon Valley, a large corporation, or anywhere else." Hwang believes that the answer to how we build innovation ecosystems is the difference between the recipe and the ingredients. Human systems become more productive the faster that the key ingredients of innovation—talent, ideas, and capital—are allowed to flow freely without following a recipe. He argues that things like diversity of talents, trust across social barriers, and "promiscuous" collaboration are the culture of his rainforest.

Hwang's new ambition is to build an organization focused on entrepreneurial rights, what he calls "the right to start." Through his experience and research, he's concluded that the unfettered right to start a new business has been repressed—that we need to change values. The Small Business Administration (SBA), for example, is about regulating investments rather than looking systemwide. "How do we unleash opportunity?" he asked. "The SBA is focused on small business rather than new business."

Hwang also mentioned his support for a local online news site devoted to tech news. Founded in May 2015, *Startland News* is a nonprofit newsroom "elevating Kansas City's innovation community of entrepreneurs, startups, creatives, makers and risk-takers through objective storytelling."[28] Not surprising, Kauffman is a major funder. Often overlooked by people outside the tech community, such news sites are important for knitting together the members of the tech community into an ecosystem. Seattle's is GeekWire, which is an essential ingredient in keeping our flywheel spinning. Other cities should consider these models as they build their flywheels.

It had been a fascinating week in the heartland, and as I headed back to Seattle, I reflected on the talent and emerging flywheels in these cities. What I saw in Oklahoma City, Tulsa, and Kansas City—commitment from local leaders, smart work centered on bottom-up support for startups, and developing talent with next-generation skills—is evidence that flywheels can take root everywhere.

~

On the flight home, I also reflected on the role of local and state governments in developing a tech economy. All three of the cities I visited had dedicated public officials working to restore their downtowns and solve local problems. I think of Mick Cornett, three-term mayor of Oklahoma City, who promoted major bond issues, which combined with private money invested over $5 billion in improvements, including a convention center, a trolley system, sports stadiums, seventy new or renovated schools, and technology projects.

Former Kansas City mayor Sly James hired a chief innovation officer in 2016 who told the *New York Times* that weaving technology into the future of his Midwestern town was a do-or-die matter. Otherwise, he said, Kansas City was in danger of becoming a "digital Rust Belt" within twenty years.[29]

But beyond the difficult work of creating a livable city, the creation and acceleration of a technology flywheel has to come mostly from the private sector.

Many states and cities have tried for some time to build their economies by chasing large businesses to locate their corporate headquarters, manufacturing plants, and servicing offices, such as call-centers, in the state. Washington State has been singularly unsuccessful in trying to lure company headquarters with financial incentives for companies to move there. Over thirty years of working on business development, I know of only one major company that moved its corporate headquarters to Seattle. That was Burlington Northern Railroad, which moved to Seattle in 1981, without state or city incentives, to break its one-hundred-year-old corporate bureaucracy centered in Saint Paul, Minnesota. After I helped them put together a legal and financial plan that freed them from their "unbreakable" J. P. Morgan 100- and 150-year bond mortgages, they morphed into Burlington Resources and moved to Fort Worth, Texas

Manufacturing plants and service offices, even if they move to the Seattle area, which is highly unlikely because of high labor costs and unionization, can easily move out of state. Even Boeing, which was founded in Seattle and concentrated most of its plane production there for its first one hundred years, did not hesitate to move its corporate headquarters to

Chicago in 2001. Likewise, after Washington State provided Boeing with billions of dollars of tax breaks to keep its manufacturing facilities local, Boeing has moved most of its operations and many of its jobs to the South and Midwest. Seattle's main efforts at economic development have been a mostly futile attempt to keep small manufacturing and warehousing within the city limits; futile because of the high costs for such businesses operating in Seattle.

Attracting a manufacturing plant or fulfillment center can be beneficial for a rural area since it often creates jobs and, in the case of Amazon and some other national companies, at higher wages than existing local wages. Amazon's two fulfillment sites in rural Mississippi employ more than one thousand employees at no lower than its national minimum wage of $15 plus benefits, compared to local minimum wages which are often set at the national minimum wage of $7.25. But it is not a flywheel strategy unless it includes jobs requiring special talent.

The Seattle area, 850 miles north of Silicon Valley, is now the home of the major engineering offices of Google, Facebook, Apple, and others. Over 140 tech companies, many of them headquartered in Silicon Valley, have established software offices in Seattle. This was not a conscious program by city or state government. The tech companies came to Seattle, attracted by the local talent and our tech flywheel, without government incentives or even awareness.

Seattle, like San Francisco, is now experiencing an outflow of tech workers and companies because of the failure of these cities to solve their civic problems and the antitech attitudes and actions of their city councils. Amazon has likely capped its growth in the number of employees within Seattle's city limits because of these issues, but a large beneficiary is the Seattle region, as Amazon has announced that it will grow offices in the nearby suburb of Bellevue to twenty-five thousand employees, where the public schools are better and the local government is more balanced in its views of the importance of technology and tech companies to jobs and economic development.

Amazon, after it withdrew its failed selection of Queens, New York, for twenty-five thousand employees, has been very public about adopting a multiple-city strategy.[30] Before Amazon's public competition to locate its second headquarters, the company had quietly conducted an internal analysis to identify all the cities in the world where it expected it could build a five-thousand-person or more tech workforce. At the time of its announcement to select Crystal City, Virginia, a suburb of Washington, DC, as its

HQ2 with a goal of hiring twenty-five thousand employees for Crystal City and five thousand for Nashville, it announced that it was selecting seventeen other cities as Amazon tech centers, including Austin, Boston, Pittsburgh, Vancouver, and Toronto.[31]

More recently, Amazon has added Dallas, Denver, Detroit, Phoenix, and San Diego to its earlier list of satellite cities. These hubs are intended to be centers of innovation that are integral parts of Amazon's mission to create software and services for its customers. Together, they already have more than twenty thousand employees. Included are jobs in Amazon Web Services, Alexa, Amazon Advertising, Amazon Fashion, OpsTech, and Amazon Fresh. Amazon HR head, Beth Galetti, said that Amazon looks forward "to helping these communities grow their emerging tech workforce."[32]

There is an enormous opportunity for cities and towns to benefit from the proliferation of tech satellite offices. In addition to Amazon, other tech companies are likewise opening tech offices across North America and globally. Apple is creating a campus in Austin which will initially consist of five thousand workers, and Uber will have three thousand in Dallas. Microsoft is opening tech hubs in Atlanta and Portland, in addition to existing offices in the Bay Area, Boston, Charlotte, and Fargo. Google plans to grow existing offices and establish new ones in Atlanta, Durham, Reston, Houston, Pittsburgh, and Portland. Salesforce has twenty-three thousand employees in Indianapolis.

What most of these cities have in abundance is livability—generally less expensive housing prices than Seattle or San Francisco and better public schools. With the explosion in remote work spurred by COVID-19 and likely to continue postpandemic, such factors will become ever more important to innovation workers. While surveys of employers and employees differ, as much as a quarter of the 160 million U.S. labor force is expected to stay fully remote in the long term, and many more are likely to work remotely a significant part of the time.

If a city wants to provide financial incentives for tech companies to locate offices in their areas, it is far better for them to invest the money in local education institutions to strengthen their tech offerings, in K–12 schools, and in civic improvements. Crystal City offered Amazon relatively small direct financial incentives, particularly compared to New York, but offered to invest millions into its local universities and schools. These are important long-term investments for Amazon and the surrounding residents.

For cities which are not likely to attract satellite offices, work on attracting talented workers who can work remotely without a local satellite office, which many companies are allowing. Richard Florida in March 2021 listed what he called "zoom towns" which are attracting remote workers, including small towns such as Boulder, Bentonville, and Tulsa, and more rural communities like Bozeman and Jackson Hole.[33] Lots more places such as Boise, Salt Lake City, and Oklahoma City could be added to these lists. Expect many tech companies to allow and recruit remote working in these places.

What most of the cities attracting tech satellite offices and tech workers have in abundance is livability—generally less expensive housing prices than Seattle or San Francisco and better public schools. And they are more welcoming to tech companies and innovation. With the explosion in remote work spurred by COVID-19 likely to continue post-pandemic, such factors will become ever more important to innovation workers.

Cities should ask themselves how best to attract talented people and engineering and other creative offices. My answer, in addition to making their cities more livable: work with businesses to speed up their economic flywheels. Launch initiatives specifically designed to appeal to remote workers, such as Tulsa's networking program for bringing remote workers together to share experiences and opportunities—and just possibly, starting companies together.

Part 3

8

Livable Cities

A Seattleite who had somehow fallen asleep in 2000 and awoke in January 2020 would think he had been transported to another city. Downtown, all but deserted after 5:00 p.m. in 2000, would be teeming with life. The South Lake Union area, previously pocked with vacant lots, cheap motels, and car dealerships, would now be home to Amazon and have been transformed into a metropolis of glittering skyscrapers anchored by trillion-dollar Amazon and four-story futuristic glass orbs known as Amazon Spheres. If our Seattleite awoke in the months before COVID-19 struck, he would have been amid hordes of tech-savvy newcomers busily streaming among their offices, downtown apartments, workout facilities, and restaurants. Just a few blocks away, the Bill & Melinda Gates Foundation, working to eradicate global poverty and disease with Microsoft-made money, has become the largest philanthropic force in the world.

Long considered a second-tier city, Seattle has clearly arrived, envied by outsiders for its dynamic, tech-driven economy, and barely bothering to hide a certain smugness at its success. In addition to its traditional technology clusters of internet, e-commerce, mobile, and enterprise software companies, Seattle has dozens of new companies built around the next generation of cutting-edge technologies. Among them, AI, cloud computing, quantum computing, and the commercial use of outer space. Biotech is booming too, with Juno Therapeutics, Sana Biotechnology, Nautilus Biotechnology, and Seattle Genetics.

Two of the three most valuable companies in the world, Amazon and Microsoft, have drawn thousands of high-skilled, well-paid workers to the region, and two of the world's wealthiest individuals, Jeff Bezos and Bill Gates, live here within a mile of each other on the Lake Washington waterfront. Fifty percent of Seattle families have incomes over $121,000; 25 percent over $200,000.[1]

This surge in creative workers and wealth has created positive ripple effects, providing a tax base and private contributions that support the arts, medical research, social welfare, schools, a new rail rapid transit system, and a $15 minimum wage with built-in escalations. Support from the tech sector together with traditional private donors allowed the Seattle Art Museum to grow a world-class collection and create a sculpture park on the waterfront. The new wealth powered huge capital drives by the University of Washington, propelling the Paul G. Allen School of Computer Science & Engineering—with two new buildings and tripling its offering of undergraduate computer science majors—to become one of the highest-ranked in the nation. Soaring retail sales and property taxes increased city revenues from $3.85 billion in 2010 to $5.9 billion in 2019.[2]

But our recently awakened Seattleite would also see thousands of homeless people, many of them living in parks or tents pitched alongside sidewalks and highways. He would sit in our traffic-congested roadways that frustrate our well-to-do and ill-serve our less affluent. He might assume that a city of highly educated, well-paid tech workers would have world-class public schools, but he would be shocked to learn that the academic disparities between middle-income white children and low-income students of color that tarred us two decades ago have only worsened. Our public schools continually fail to lift children from families of limited means and equip them with the skills to participate in tech-driven economies, such that most disadvantaged children have no hope of employment in Seattle's high-earning tech jobs.

The growth of our innovation economy has not been sufficient to solve our social problems. The economic dynamism of Seattle and resulting individual wealth are real, but also real and highly visible are the city's unsolved—and worsening—problems.

The growth in the number of high-paid workers has exacerbated some of these social problems. Many workers, such as teachers, police, and service providers, can no longer afford to live in the inner city, forcing them to live farther from their jobs and endure traffic that wastes their time and pollutes our air. An average home here now goes for $856,000. Homeownership, once a cornerstone of the American dream, is out of range for many.

Homelessness, meanwhile, is tragically being experienced by thousands. It increasingly infringes on daily life downtown and in our neighborhoods. Hypodermic needles litter the grounds of our parks. Higher housing costs push some people into homelessness, but much of homelessness is driven more by drug addiction and insufficient mental health services.

A direct result of Seattle's booming economy is traffic congestion, worsened by a transit system that relies primarily on buses. We often can travel within the downtown faster on foot than by driving—while pedestrians and cyclists must endure the heavy pollution from diesel buses. The less affluent are forced to spend hours each day on multiple bus transfers to reach work and needed services.

The disparity in education was highlighted by the COVID-19 crisis when the superintendent of Seattle Public Schools closed all the schools with no immediate plan to provide remote learning because of equity concerns that not all students had access to computers and sufficient bandwidth. When the Seattle school district reopened with remote learning, low participation rates in its low-income schools further aggravated the disparity.

Many suburban and rural schools, as well as public charter schools and private schools, were mostly successful in transitioning to remote learning. But private schools are expensive and cannot meet demand, and public charter schools, which are free and open to all, often have long waiting lists caused by state legislatures that impose limits on their growth. Even though this option often is sought by minority parents, many progressives oppose their growth. Inequities only grew when some public-school families decided to pay teachers privately to work with small pods of students.

As I write this in April of 2021, Seattle's downtown is again deserted, and not just after five o'clock. With the impacts of COVID-19 and Black Lives Matters protests, office tenants are working remotely, and many of the residents have fled to second or even their parent's homes outside the city. Over one hundred retail stores are fronted not by display windows but by plywood to protect against violence. Many restaurants are closed, an unknown number permanently. Tents that crowd the streets and aggressive panhandling and crime make people fearful and deter suburban residents from their trips downtown for shopping, restaurants, and entertainment.

Members of city councils in Seattle and San Francisco have become increasingly radicalized and, instead of looking forward to new models of working with business, are hostile to new technologies and tech companies, adopting regulations restricting the introduction of new technologies and imposing new taxes on companies and their employees. At the same time, the city councils are failing to solve the problems of homelessness, public safety, transportation, and education.

For the first time, I worry that our inability to solve the social problems of our city is becoming a drag on its continued success. Will we be able to attract and retain the kinds of creative, entrepreneurial workers we need to sustain our prosperity? Amazon's announced plans for locating twenty-five thousand workers not in Seattle but across Lake Washington in downtown Bellevue suggest that these worries are warranted.[3]

The question has been thrown into even sharper relief by the COVID-19 effect, which caused many workers logging in from home to realize the benefit of not having to fight with Seattle's traffic congestion and public safety challenges. Not surprisingly, smaller cities like Tacoma, Tulsa, or Burlington, Vermont, are seeing a surge of newcomers, and their economic flywheels are poised to take off.

Small- and midsized cities are not without social challenges, including drug abuse, poverty, and histories of discriminatory policies, but it is in economically successful tech cities that the problems appear more deep-seated and unsolvable. The economic flywheel has created jobs and wealth, but it has not solved the social challenges of cities.

What we and other tech centers need now is an additional kind of flywheel, one targeting social challenges with the same kind of energy around innovation that we brought to our economy—I call it a livability flywheel. And it needn't depend on fancy tools or lots of money. It requires not rhetoric but practical programs targeted at each problem, using social innovation as well as economic innovation. In chapter 9, you'll see an example of the way one small city, Eugene, Oregon, has deployed medics and crisis workers to address problems traditionally turned over to the police.

As I envision it, the livability flywheel (figure 8.1) starts with public and private investments in the cityscape and services that create the kinds of vital downtown areas that happened in Seattle and I saw in Tulsa and Oklahoma City, with waterfront parks and restored and new buildings designed for mixed-use of residential, business, and recreation. These are the same types of improvements that help create the economic flywheel.

COMMUNITY
LEADERSHIP
(Mayors, CEOs, Civic)

LIVABILITY

LIVABLE
CITY
(Downtown,
Parks, Housing,
Education, Jobs)

BUSINESSES
(Startups,
Established)

TALENTED
RESIDENTS

Figure 8.1
Livability flywheel *Source*: provided by the author.

These improvements keep and attract the talented people the city needs to grow and continue to improve. They provide civic and business leadership, create jobs, and in turn, make the city more livable. Even more than the economic flywheel, the livability flywheel depends on enlightened political and civic leadership, particularly a strong mayor who can rally the disparate interests among the electorate to support practical solutions. The livability flywheel uses wealth created by the economic flywheel to invest in solutions with an innovation mindset.

The difficulty of achieving these goals was brought home to all of us by the Black Lives Matter protests of 2020, which dramatically highlighted many serious issues in our cities of widespread discrimination, police treatment of minorities, excessive police tactics, housing segregation, unequal public schooling, discrimination in hiring and promotion, and disparities in income. There were marches of thousands of people in over one hundred cities, large and small, lasting in some places for months, showing widespread support for changes. Unfortunately, rioting, looting, and violence by an extremist minority took advantage of the protests and detracted from the serious discussions for reform, and even drowned out the concerns of Black leaders that law and order are also important to their communities.

All whites are not racist, but the need for eliminating racism in many of our schools, businesses, and other institutions is inarguable. Local governments, nonprofits, and companies have been prominent in pledging support and action, but I question whether most of them are taking the steps necessary to solve our most important long-term issues, such as providing the equal education for Black and other disadvantaged children that is essential for creating the opportunities for minorities to achieve financial and social equality. Newly launched programs to reform police departments and to recruit, mentor, and promote people of color in businesses and government will help—but a bigger effort by everyone is required.

Growing up and living most of my life in progressive Seattle, I, like many whites in our community, believed we lived in a nonracist enclave. We knew about the discrimination in the South and supported the marches and federal actions of the 1960s to provide equal rights to Blacks. But we considered it a Southern problem (even as Seattle practiced residential redlining into the 1970s). But, other than our occasional actions to eliminate the exclusionary policies against women and Black people at Seattle business and social clubs, we were not particularly aware that all Black people, of whatever economic or social achievement, experienced racism in our community. Nor did we undertake a wide-reaching reform effort to improve the lives of people of color.

I must admit that the first time that I became fully aware of the extent of discrimination in Seattle and generally in the business and academic communities across America was when a highly regarded Black lawyer, Steve Graham, who had worked for me while we were lawyers at Perkins Coie, self-published a book through Amazon in 2017: *Invisible Ink: Navigating Racism in Corporate America.*[4] Amazon's summary is tellingly shocking.

> *Invisible Ink* recounts Graham's experiences with bias and racism in corporate America. Unlike racially motivated violence or overt bigotry, racial bias in the business world is usually subtle, often going undetected unless coaxed to the surface. Such racism is insidious and deeply ingrained in corporate America. Succeeding means battling against prejudice on a daily basis—all while white colleagues maintain

racial bias doesn't exist or is of no consequence, dismissing attempts to confront prejudice as "playing the race card."[5]

If published today, Graham's book would be a blockbuster. In 2017, it was hardly noticed, not deserving a review in the *New York Times* nor any other national publication.

I was surprised and initially proud when Graham recounted how he had received help from several white individuals:

> No one does it alone. Throughout history, there have always been people of goodwill whose minds and hearts are more open and whose level of courage is much greater than that of most who are willing to step in and work against the grain to make a difference. I will be forever indebted to the following individuals who stood beside me when others did not: . . . Tom Alberg, the Perkins Coie partner who took me under his wing and stood against my detractors, cutting a trip to New York short so that he could return to Seattle and argue for my partnership in front of the executive committee, some members of which had vowed never to make me a partner.[6]

But then I read this from Graham's book and realized I had acted out of an instinct of fairness but didn't realize the greater context of discrimination that existed.

> Even those who do care don't really understand. This is all played out in an environment where we are subjected to a debilitating undercurrent of bias that too many, on both sides of the divide, pretend does not exist. . . .
>
> The point of this book is not that the world is an awful place where things never go right but that institutional racism is a virus that is alive and well and needs to be eradicated if fundamental fairness is to be achieved. Black lives matter, and we must take issue and demand change, whether these lives are literally snuffed out in the blink of an eye or figuratively snuffed out in the polite confines of corporate America.[7]

Many of us once dreamed that Seattle's technology economy would lead a renaissance of urban life in America, with a growing creative class welcoming a diversity of immigrants, races, and sexual orientations. We imagined that technology and the wealth it generated could provide shared economic prosperity for all—including, along the way, new solutions for long-persisting public problems. Something very different has happened. There is no logical reason why a city full of innovative problem-solvers cannot solve these social riddles. But the problems have now grown to such a scale that I wonder if creative, forward-thinking residents will weary of our ongoing failure to make headway.

Will major cities solve their homeless crises? Will public safety be restored? Is improving public K–12 education in big cities an unsolvable problem? Will office buildings fill with workers or will many of them choose to work remotely? Will downtowns return to being the supercharged places they were with restaurants, shops, and people?

In the following chapters, I suggest how local governments and businesses can work together to solve these problems.

9

Public Safety and Privacy

In early 2021, Seattle's downtown was bleak. Primarily caused by COVID-19 but also homelessness and civic unrest, the city's downtown was mostly deserted—and not just after 5 p.m. Office workers were logging in remotely, and residents were fearful of coming downtown because of an increase in crime. This was fueled by open drug dealing and the increasing number of mentally ill homeless people on our streets, many living in tents on sidewalks. Apartment and condo residents had fled to parents' or rural second homes. More than a hundred retail stores were now fronted not with glittering window displays but plywood—to protect against violence. Nordstrom's flagship store, with its corporate offices above, stood out with its storefront windows welcoming, but shoppers were sparse.

Visitors and residents were accosted on city streets, and car windows were smashed for grabs of purses and laptops. One of Seattle's councilpersons introduced legislation to legalize misdemeanor crimes—like shoplifting, vehicle prowls, and threatening someone with a gun—if perpetrators show symptoms of substance abuse or mental disorder or that it was related to poverty.

In family-centered neighborhoods like Ballard, where I grew up, feces and hypodermic needles littered local parks, and residents feared thievery, assault, and rape.[1] Children were not allowed to leave their yards.

An explosion of understandable citizen anger over long-persisting police inequities was hijacked during the Black Lives Matter peaceful demonstrations by an extremist minority who broke windows, graffitied store entranceways, and set a police cruiser aflame. Night after night, Seattle's street violence made headlines. Protests lasted for weeks and at times involved thousands of people. Lawlessness was also much in evidence across the country. In Minneapolis, Atlanta, Portland, and many other cities, rioters smashed shop windows, ransacked shelves, and attacked government buildings, detracting from a serious and necessary discussion around police reform.

The impact of COVID-19 on our cities, even though severe, will dissipate with our success in bringing it under control, but I fear that cities such as Seattle and San Francisco will find their economic futures jeopardized unless they can make substantial progress post-COVID-19 eliminating homelessness, improving public safety, reducing police abuses of minorities, and providing more opportunities for minorities to move up the economic scale through improvements in education and elimination of discrimination in schools and employment.

This is an imposing agenda. But if we find ways for businesses and governments to work together to solve these problems, I believe it is possible to restore downtowns as hives of activity with workers, shoppers, restaurants, shops, and throngs of tourists, and the economic flywheel can generate businesses, jobs, and wealth for everyone's benefit.

Public Safety

How best to combat criminal behavior and make our cities safe is not easily answered—particularly in the current pressure cooker of emotion created by marches and riots generated by police violence and killings.

Public safety was a major problem in our leading tech and other cities before COVID-19 and protests and riots. Public officials and commentators cite conflicting statistics as to whether overall crime and murders are up, down, or flat, but ask downtown workers and city residents and they will tell you that crime is a serious problem in their neighborhoods with assaults, robberies, car break-ins, and drugs. Between 2014 and 2018, aggravated assaults in downtown Seattle rose 60 percent, and burglaries were up 64 percent. Murders are up significantly in most major cities. Equally serious, the amount of unsolved crime and murders are up, which creates a negative flywheel. In middle-class neighborhoods, crimes against people—rape, robbery, and aggravated assault—increased between 64 and 86 percent.[2] A common complaint in Seattle is that "I cannot allow my children to go to the park alone."

Part of the increase is due to more people living on the streets with addiction and mental health problems, aggravated by lack of enough low-cost housing, and failures in our judicial system, incarceration facilities, and mental health and drug treatment and facilities. With the lack of mental health facilities, the choice for the mentally ill is either living on the streets or jail.[3]

The Seattle City Council banned the police from using pepper spray and other crowd control tools, and the police chief, Carmen Best, informed the council in late July that as a result, police officers could not preserve public safety when large crowds become violent. Increasingly, business associations and neighborhood groups hired security forces to patrol their neighborhoods, further exacerbating the divide with poorer neighborhoods that have the same or worse crime problems.

There is strong evidence that police departments need to be reformed, but the rush by city councils to pass resolutions defunding police departments, or in Seattle's case proposing to cut police budgets by 50 percent,[4] were convenient political slogans. Needed reform is hard work and will require smart restructuring of our public safety systems. Improving public safety likely will require more money, not less, to upgrade officer training, strengthen mental health interventions, and bolster youth outreach while maintaining resources to fight crime. Activities such as jaywalking should be decriminalized, and police functions such as chasing deer from residential areas should be dropped or transferred to nonuniform city workers. The handling of issues such as minor drug use and selling and prostitution are more controversial but in many communities are being liberalized.

The police disciplinary process requires reform in many cities, as it currently protects too many officers from removal. In Minneapolis, since 2012, there have been 2,600 civilian complaints against police, but only twelve resulted in any form of discipline.[5] Even though police unions have vigorously defended disciplinary processes, they often provide a needed balance to city councils' attempts to strip police officers of tools to fight crime and protect themselves.

Reforms in training and disciplinary processes also need to be combined with the expanded use of new technologies in nonabusive ways. New technologies in the forms of improved video cameras, photo ID, large digital databases, artificial intelligence, and DNA matching have great potential to improve significantly our ability to deter, identify, and prosecute criminals.

A promising reform proposal by a Seattle councilperson, which appears to have worked in Eugene, Oregon, provides for a first-response unit independent from the police that responds to 911 calls that do not require armed police. For about twenty-four thousand emergency calls in Eugene annually (20 percent of the total calls), unarmed teams composed of a medic and a mental health crisis worker respond. These professionals offer conflict resolution, housing referrals, first-aid, and immediate transportation to shelters. In only about 250 of the calls was it necessary to request police assistance.[6]

In Eugene, the program has resulted in average savings of $8.5 million in police costs each year[7]—or nearly $30 million since 2015—in addition to $14 million in annual ambulance and emergency room costs.[8] Denver recently created a similar program. Whether Seattle is capable of creating and successfully managing a similar program remains to be seen.

Privacy and Public Safety

Public Surveillance

The right not to be spied upon is fundamental to our understanding of liberty. But does this apply every place in all circumstances? Obviously in our bedrooms and throughout our houses. What about public streets, alleyways, transit stations, and buses? What about shopping centers and elevators? What about if we enter the area of your neighbor's front porch or side yards? When is privacy trumped by the needs of public safety?

Grainy office building videos watched by security personnel on small 1960s-equivalent TV screens have been replaced by high-resolution cameras and screens with night vision capabilities and supersharp screens monitored remotely by humans or AI systems that send alerts based on motion or noise. Video systems are being installed worldwide by governments and police departments, businesses and individuals, but the full deployment of these technologies depends on convincing citizens and politicians that their use will not unreasonably infringe on personal liberties.

For many people, the idea of cameras watching our every move feels too close to the constant surveillance in *Brave New World* and *1984*. China shows us the real-life abuse of this technology by an Orwellian government.

Reportedly, 170 million video cameras are watching China's citizens. This network, paired with China's political dictatorship, unfettered by a legal system that protects personal privacy, has created the most advanced system in the world for identifying and tracking its people. In addition to using video to deter and catch criminals, Chinese authorities are using them for somewhat benign purposes such as shaming jaywalkers by publishing their names.[9] More serious is their use to track and punish protestors and critics. The system can even be used to punish those who associate with such persons by denying them tickets to trains and government benefits.

Beyond their comprehensive video surveillance system, the Chinese government is developing "records on each citizen's political persuasions, comments, associations, and even consumer habits. The new social credit system under development will consolidate reams of records from private companies and government bureaucracies into a single 'citizen score' for each Chinese citizen."[10] Combined with near real-time searching, the Chinese government is fast becoming the ultimate eye-in-the-sky police state.[11]

You don't have to look to China to see the widespread adoption of video surveillance. You can find it in leading Western cities such as London, Boston, and New York. There are "approximately 420,000 closed-circuit television cameras in London, more than in any other city except Beijing, equaling about 48 cameras per 1,000 people." Purportedly, sensitive enough to scan airport crowds for a single tattoo, London investigators combed through millions of images from London's cameras to identify and track the movements of the al Qaeda-inspired bombers who attacked a bus and three subways in 2005. Investigators also used video images to identify the killers in the 2018 poisoning of Sergei Skripal, the former Russian spy.[12]

In Boston, a similarly extensive network of surveillance cameras helped to pick out, amid a crowd of thousands, the young man who bombed the 2013 Boston Marathon crowd.[13] In Newark, New Jersey, people sitting at home in their living rooms can live stream footage from police surveillance cameras.

The value of this technology to pick out criminals is obvious. Cities deploying video cameras believe that they also deter criminals.[14] Yet video surveillance in cities has begun to attract criticism. San Francisco banned all use of video surveillance by police in 2019. Although Seattle has not formally banned its use, it has set up an elaborate approval process that has yet to be used. And, previously, in 2013, twenty-eight cameras installed by the police department in high-crime areas of Seattle with a grant from Homeland Security were never activated and have been removed.

Although police in Seattle are barred from using video cameras other than for capturing traffic infractions and as body cams, the benefits of monitoring public places with video cameras were obvious when the cameras of the Seattle public transit authority showed a video of a gunman emerging from a subway station carrying a pistol in his left hand that he had just used to shoot two people on the subway platform in the station below Nordstrom's flagship store. The picture allowed the police to identify and capture the gunman after the *Seattle Times* published the photo on its front page.

The trade-off between the desire for personal privacy and public safety was illustrated during the BLM protests in June 2020, when Seattle's mayor wrestled with the issue of police body cameras. Several years earlier, in response to the demands of civil rights leaders, all Seattle police officers were required to wear body cameras to monitor—and presumably deter—excessive use of force during arrests. But privacy advocates complained. So Seattle required that the body cameras be in off mode most of the time, only switched on when officers were detaining a suspect. After the killing of George Floyd in Minneapolis sparked demonstrations nationwide, Seattle flipped yet again; now protest leaders demanded that police body cams be in on-mode all the time. A double-edged sword, to be sure, as the cameras were then able to capture footage of nonviolent marchers.

Private Surveillance

Even if a local government refrains from deploying video cameras in public places, private companies and homeowners are installing video cameras for their premises that also provide views of public spaces and streets. The New York Police Department's Domain Awareness System consists of over 9,000 public and private video cameras.[15]

In San Francisco, a tech entrepreneur has financed the installation of a network of more than one thousand high resolution video cameras over a 135-block area that links Union Square to Japantown, Fisherman's Wharf, the Tenderloin, and Russian Hill.[16] Although police don't have direct access to the videos, the neighborhood groups that operate the network will provide footage to them upon request.

Homeowners are rapidly installing video cameras outside and inside their houses. Ring.com, a company Amazon purchased in 2018 for $1 billion, has developed low-cost, internet-connected video cameras integrated

with doorbells that can view the porch and the street beyond. Other connected cameras can be installed to view the entire periphery of the property or indoors, which can instantly notify a security company with a short
video when someone is seen by the camera or prompted by the sound of
glass breaking or a barking dog. Ring's app Neighbors can share videos of
suspected criminal activity with neighbors and the police. Though critics
have raised the specter of vigilantism law enforcement, agencies in over
fifty cities are integrating it into their crime prevention and capture programs.[17] Many other companies are offering similar cameras and systems:
"The partnerships let police request the video recorded by homeowners'
cameras within a specific time and area, helping officers see footage from
the company's millions of Internet-connected cameras installed nationwide, the company said."[18]

Advanced capabilities are being developed that will be able to distinguish family members from intruders based on body shapes, even if facial
views are not seen.

A recent technology advancement is the use of video cameras on automated aerial drones. Police departments have begun to use camera drones
that are easier to deploy and less costly than helicopters to visit a crime
scene and chase after a fleeing criminal.

> The latest drone technology—mirroring technology that powers self-
> driving cars—has the power to transform everyday policing, just as
> it can transform package delivery, building inspections and military
> reconnaissance. Rather than spending tens of millions of dollars on
> large helicopters and pilots, even small police forces could operate
> tiny autonomous drones for a relative pittance.[19]

Amazon has introduced a flying camera drone for homes and businesses
that when activated by an alert can fly around the inside of your home or
business transmitting what it sees to your Ring app before returning to its
charging dock. It costs $250.

Will cities try to prohibit personal camera systems which some are
suggesting will create a "citizen-enabled surveillance state?" People may be
squeamish about these systems, but for many, it helps them feel safer.[20]

Combining facial recognition and AI with video surveillance significantly increases its power to identify criminals. Amazon Web Services offers
Rekognition, an AI-enabled service that can identify suspects by matching
their photos against a database of thousands, or even millions, of facial

images, scrolling through them with a speed and precision impossible for any human detective. Some agencies have deployed it to identify and arrest criminal suspects in crowded public events, such as football games. Amazon says that it has "seen the technology used to prevent human trafficking, reunite missing children with their parents, improve the physical security of a facility by automating access, and moderate offensive and illegal imagery posted online for removal."[21] Schools are using it to boost security.[22]

In 2018, a test of Rekognition by a MIT lab concluded that the service discriminated by misidentifying people, particularly minorities and women. Amazon blamed user error, claiming that MIT had created faulty results by basing its test on a 70 percent probable match, whereas AWS recommends to police that they use a 98 percent standard and verify it with a human review.[23] The press frequently cites the MIT study as proof that Rekognition is inaccurate and discriminatory but almost always omits Amazon's response.

In 2019, without admitting its software discriminates, Amazon announced that it was suspending its use by police (except for human trafficking cases and finding missing children) for one year to give Congress time to adopt legislation regulating its use. Microsoft and IBM promptly followed suit, with IBM going so far as to say it was stopping all selling of recognition technology. The ACLU responded that these actions were not enough, and all recognition technology should be banned: "No one should be unjustly harassed by this technology, nor should anyone have to worry about their face being scanned, stored, and sold by companies."[24] Others demand that tech companies stop selling all technologies to police departments.

Some members of Congress support limiting the use of smart video, with Representative Jim Jordan of Ohio, the top Republican on the House Judiciary Committee, comparing the technology to George Orwell's *1984* and a threat to free speech and privacy. A school in Sweden was fined after using facial recognition to keep attendance.[25]

In 2020, the Washington State legislature authorized the use of facial recognition technologies by the government but imposed strict limits requiring advance community notice, a warrant or court order, independent third-party testing for accuracy, or bias and human review before any government action. Although endorsed by Microsoft, it did not enjoy universal business backing and the ACLU opposed it in favor of a bill to ban surveillance video.

Cities are also imposing restrictions. Oakland, San Francisco, and Boston have outright barred government use of facial recognition technologies.

Portland has prohibited private businesses from using it in public areas of stores, banks, entertainment venues, and sports stadiums.

Amazon has called for federal legislation on facial recognition "while also allowing for continued innovation and practical application of the technology."[26] Absent federal preemption, limits on local use of video surveillance and facial recognition will be worked out by local governments, with some imposing strict limits and others allowing self-regulation. At the same time, advocates of stricter restrictions are marshaling their efforts at federal and state levels.[27]

As artificial intelligence improves, its algorithms will produce more accurate identifications than humans can by removing the possibility of unconscious bias. Indeed, facial recognition technology received an unexpected endorsement from privacy groups that have found it useful for identifying rogue police officers at protests.[28]

Privacy and Personal Information

Routinely, people insist that they want their personal information protected at all costs. But when faced with a practical choice of giving up some privacy to allow for video surveillance that could catch a rapist, kidnapper, thief, or murderer, most are willing to sacrifice quite a lot of privacy. We even sacrifice privacy so that advertising can be better directed to us or so that our friends can view our photos. "The evidence suggests that, in fact, we love our devices as much as ever."[29]

Nonetheless, there is an international outcry against tech companies using and selling our personal information. What has been called the most comprehensive privacy law in the United States was adopted by the California legislature and then strengthened by a 56 percent public vote margin in November 2020 that gives consumers rights to control what information companies collect on them and how it is used. California residents can demand that companies disclose all information that is collected on them, request a copy of that information, require the deletion of their data upon request, and prohibit the sale of their information via a "do not sell" link on the company's website.

Businesses are responding to these attacks in part by seeking legislation from state legislatures and Congress that protect civil rights while also allowing for continued innovation and practical application of the technology but preempt local regulation. Brad Smith, president of Microsoft, advocates federal legislation regulating privacy.[30]

Genealogy and DNA

What began as family trees built and stored by people on sites like Ancestry .com has become a powerful crime-fighting tool when people submit their DNA from saliva samples to find related persons, such as an unknown biological parent or a distant relative.

Police can upload DNA found at a crime scene and find related relatives who have submitted DNA samples. Even if the suspect has not uploaded samples, the family tree can be completed from public records such as wedding announcements and obituaries. The suspect is somewhere on this family tree, which greatly narrows the search by police.

Crime-scene DNA and a family tree aren't enough to arrest or convict someone or even obtain a search warrant. Police therefore must collect fresh DNA from the suspect to confirm the genetic match. Nonetheless, this is a powerful tool, and many cold cases are expected to be solved. Over forty cases have been solved so far, including cases more than fifty years old.[31]

GEDMatch contains genetic data from more than 1.2 million individuals and is adding one thousand to two thousand genetic profiles each day. Family Tree has provided the FBI with more than a million genetic profiles. A geneticist at MyHeritage estimates that profiles from three million Americans of European descent could identify 90 percent of people.[32]

The use of this data has raised privacy concerns because some unknown parents and relatives may prefer to remain unknown, and in some cases, people who are uploading DNA data and family tree information may be unintentionally incriminating a relative.

Location Data

During the COVID-19 pandemic, many countries promoted the use of mobile phone tracking and tracing apps to fight the virus, setting aside privacy concerns. China and South Korea even required it.[33] During the early months of the outbreak, the chair of the expert panel advising the World Health Organization publicly said, "You need to identify and stop discrete outbreaks, and then do rigorous contact tracing,"[34] But though this advice was repeated across the world by numerous health officials and as results in early adopting countries demonstrated how contact tracing was protecting human health and dramatically curbing the spread of COVID-19, privacy concerns in the United States caused approval of phone-based apps by states to be exceedingly slow.

To overcome privacy objections for location-based apps that match people who were together, Google and Apple jointly developed software for an app that uses Bluetooth to alert people who have been in contact with an affected person without identifying the person or providing information on locations to a central database. Even though no information leaves a person's phone, most governments and businesses in the United States only began to endorse the apps seven months after COVID-19 was identified as a pandemic and have not aggressively promoted it, likely resulting in thousands of unnecessary deaths.[35] Likewise, large employers and unions generally did not make significant efforts to promote the apps' use.

∼

Instead, cities and states tried to set up tracking programs by hiring thousands of callers who attempted to contact affected people by phone to ask them who they had been in contact with. With the obvious problems of wrong numbers, people not answering their phones, lack of knowledge of names who were at places of contact, and faulty memory, these human-based efforts were unable to slow the spread of the virus. As a front-page headline in the New York Times noted, "Contact Tracing Has Largely Failed in Many States.": "There is little appetite in the United States for intrusive technology, such as electronic bracelets or obligatory phone GPS signals, that has worked well for contact tracing in parts of Asia."[36]

New York City, declining to use cell phone technologies, was employing three thousand callers. Three months into their calling program "only 42 percent of infected people [reached] provided the tracers a single contact they might have exposed, a level that epidemiologists consider too low for the program to be broadly effective." Twenty-five percent of those called in Maryland didn't answer their phones. The upsurge in COVID-19 cases in October further exposed the inadequacy of not using tracking technologies, as human-based tracing became mostly impossible.[37]

California Governor Gavin Newsom's office made use of location data from Foursquare Labs Inc. to figure out if beaches were getting too crowded; when the state discovered they were, it tightened its rules. In Denver, the Tri-County Health Department monitored counties where the population on average tends to stray more than 330 feet from home, using data from Cuebiq Inc. Researchers at the University of Texas in San Antonio used movement data from the geolocation firm SafeGraph to guide city officials on the best strategies for getting residents back to work. Much of the location data was coming from location-tracking firms in the ad-tech industry, which otherwise has been criticized for its use of location data.

Public Data

Cities, too, are generating enormous databases, including real estate ownership, transactions and values, voter registration, elections voted in and personal contributions, traffic information, and water usage, most of which are available to the public. Add to this the deployment of sensors and mobile apps that gather data on traffic jams and city services. Toronto[38] and Portland[39] are examples of where this has raised privacy concerns. Tech companies are often hired by cities and other companies to help them make use of this data.[40]

Sometimes it is possible to determine information on a particular individual from collections of assumed anonymous data.[41] "Despite the privacy-safe assurances, data privacy and security experts question whether mobile location data, even when deidentified, is safe from re-identification if leaked, hacked or obtained by law enforcement." It should be no surprise that political campaigns are using tracking data to categorize voters.[42]

On the other hand, innovative uses of data can help us uncover problems in our criminal justice system that need attention. For example, a group of citizens in Seattle shined a spotlight on the catch-and-release handling of the hardcore criminal offenders. They published data in 2019 that analyzed judicial records showing that the one hundred most prolific offenders were involved in nearly 3,600 criminal cases. This finding could have been used to redirect police attention or create new interventions for the identified population, but so far our political leaders have not used this illuminating use of data has to make changes to our criminal justice systems.[43] It takes action by these leaders to implement change, which too often is lacking.

COVID-19 raised new issues of privacy. For example, under federal HIPAA laws, an employer cannot disclose the name of an employee who tested positive even to fellow employees for whom such information could be valuable in determining whether to seek a COVID-19 test or to self-quarantine. As described above, fears of being criticized by privacy advocates have caused political leaders in many cities and states in the

United States to refrain from advocating the use of tracing and tracking mobile phone technologies to control the spread of COVID-19.

The use of AI in law enforcement, including smart video cameras, automated drones, and the mining of personal and governmental databases, will give us enormous powers to improve law enforcement for our safety. But technologies and data can be used to threaten or enlighten. It takes smart government leadership to determine which will be the case. Even if AI can make better judgments than humans and remove human bias, this is not to say that AI should make judgments as to what level of privacy is acceptable in our society. Those kinds of policy decisions must be left to people.

10

Homelessness and PreK–12 Education

Homelessness and Affordable Housing

It is an uncomfortable truth that in the Seattle area, where half of all families post annual incomes above $102,500, some twelve thousand people live on the streets or in shelters. At the last official count in January 2020, an estimated 3,700 were living "unsheltered." Visibly, the number of tents on streets in Seattle significantly increased during 2020. One survey reported a 50 percent increase in tents in parks, sidewalks, and other locations.[1] Walking the streets of my city you can visit the tech-saturated South Lake Union downtown neighborhood, marveling at the Amazon biospheres where coders meet amid exotic plants culled from across the world. The world's largest philanthropic organization, the Bill & Melinda Gates Foundation, headquartered a few blocks away, is eradicating global poverty and disease with Microsoft-made money. But you must also sidestep unhoused people living in tents pitched on the sidewalks—their belongings heaped in shopping carts. Thousands more are living in city parks, beneath overpasses, and alongside our highways—a bitter irony and human tragedy for a city that has long seen itself as a bastion of progressive values.

The issue of removing tents in Seattle from sidewalks, parks, and playgrounds highlights the cultural and political divide in Seattle and other cities. As students began to return to classrooms in late April 2021, many parents asked the city to remove the tents from nearby playgrounds. Members of the Seattle School Board responded: "Our students deserve to see the adults in their lives behave compassionately . . . we demand sweeps never be used on school grounds, adjacent or elsewhere in this City."[2]

Homelessness is a significant and growing problem in many of our large cities. New York has seventy-eight thousand homeless, Los Angeles fifty-six thousand, and the Bay Area thirty-three thousand.[3] Major causes are the failure of our local governments to provide adequate mental health, drug and alcohol treatments, and housing for them as well as lax enforcement of laws and prosecution of crimes against people and property. A contributing cause also is the upward pressure on rents and gentrification of neighborhoods caused by the increased number of people who want to work and live in our cities, mainly related to the tech economic boom.

Fifteen years ago, a major effort was launched in Seattle by government agencies, churches, nonprofits, businesses, and citizens to "end homelessness in Seattle," but today we are nowhere close to a solution. Our city, despite government efforts and its surging wealth, now holds the dubious distinction of having the third-highest per capita number of homeless men, women, and children in the country.[4]

Seattle officials have called this a crisis and vowed, again and again, to eradicate it. Yet of the dozens of housing programs and services Seattle has funded, a third have been judged by consultants hired by the Seattle City Council to be ineffective at helping people make permanent changes. In response, the council defunded several of the programs—but in the face of intense political lobbying by the affected organizations, the council reversed itself and restored the funding of these agencies.[5] In another controversial action, the council recently voted to defund the city's Navigation program which uses police and outreach workers to clear homeless encampments and refer people on the streets to shelters.[6] Denver, Bakersfield-Kern County in California, and other municipalities meanwhile have made far more progress in providing housing and treatment for those in need.

Another problem is the Seattle way—appointing large committees to make recommendations or decisions—which often gets in the way of taking decisive actions. An example is that the city and county recently merged their homeless programs with a $100 million budget, but the convoluted governance structure of a twelve-member governing committee and a thirteen-member implementation board[7] likely will limit its ability to advance effective solutions.[8] A better structure for the homeless crisis might be the appointment of a homeless czar to make decisions.

I am convinced that these problems can be solved—not by clinging to legacy programs that have shown few positive outcomes but by effective political leadership pursuing innovative solutions, with strong backing by businesses and nonprofits. Seattle, among other cities, is not pursuing well-known programs for solving them, as most of our present city councilpersons appear to prefer rhetoric and politics to practical solutions.

Tim Burgess, a well-regarded former city councilperson and interim mayor, has proposed a practical set of steps that could put us on a path to eliminating homelessness.[9] These include a prioritized removal of illegal tented encampments on our streets and parks with interim accommodations, tiny houses, and rented hotel rooms, with a longer-term commitment to permanent housing and on-demand treatment for substance abuse disorder and mental health issues rather than scheduling appointments.

Many experts believe that the number one need for solving homelessness is building sufficient housing for them and that a "housing-first" policy should be pursued to reduce homelessness—meaning that there should be no preconditions such as being substance-free to enter homeless housing.[10] With stable housing, then government service agencies can work to improve the residents' lives. A hard core of as many as two thousand has mental or physical health problems that make it likely that they will have to live permanently in dedicated housing.[11] For most of the others, homeless housing can be temporary, and with sufficient supporting services they can reenter the economy and live productive lives.

The public health crisis created by COVID-19 forced Seattle to take actions that provided evidence in support of a housing-first policy. The county rented rooms from hotels around the region—empty because of the pandemic—as temporary private residences for nearly nine hundred people who had been living on the streets.[12] In the nearby suburban city of Renton, an unoccupied Red Lion Inn became home for two hundred homeless. "For the first time in many of their memories, each resident had a room and bathroom of their own," said Daniel Malone, head of Seattle's Downtown Emergency Services Center. "Almost overnight, people's lives improved."[13]

The effort was conceived as an emergency stopgap to stem the pandemic, but its success has begun to inspire larger visions. Hotel residents reported better physical and mental health and an improved ability to focus on long-term goals like finding jobs and permanent housing, rather than scrambling to survive day-to-day.[14] County officials are now trying to determine whether the model might be expanded to provide temporary homes for two thousand people and have begun to purchase small hotels that have been adversely affected financially by the pandemic. Other cities have launched similar purchasing programs.[15]

Providing housing and mental health and other services for all the homeless takes considerable money since rents need to be free or highly subsidized. Some of the homeless who have jobs can pay all or part of a low rent. Others receive federal and state monthly payments—a percentage of which is applied to rent. Through combined nonprofit, business, and government efforts we are spending about $300 million annually to build new homeless housing in the Seattle area, but according to a McKinsey study, additional construction funding of "between $4.5 and $11 billion over 10 years" is needed to build the subsidized units necessary to fully house the homeless in the Seattle area.[16]

Sounds like a lot of money, but isn't this something that our companies and nonprofits could commit to accomplishing over the next few years? Challenge Seattle could take the lead in creating support for such a fund.

In fact, in 2019, Chris Gregoire, former governor of Washington and CEO of Challenge Seattle, commissioned a Boston Consulting Group (BCG) study of low-income housing needs in the Seattle area and proposed at a dinner meeting, that I attended, of the Challenge Seattle CEOs that the companies create a $1 billion loan fund to develop "worker housing." BCG quantified the total need for low-income housing as $10 billion compared to estimates of $3 billion necessary for homeless housing.[17] The CEOs recognized the problem of the lack of low-income housing, but some of the Seattle CEOs suggested that their companies and employees preferred to prioritize homeless housing. One tech CEO said that if he proposed at an all-hands meeting to prioritize worker housing rather than for homeless housing, he "would be booed off the stage." Other CEOs suggested that we might begin with some pilot projects. There was no consensus to move forward at that time.

A few months later, Microsoft—likely influenced by the study— announced its program to provide low-interest loans aggregating $475 million to developers of low- and middle-income housing but only $25 million to support homeless housing.[18] Microsoft and Gregoire worked together to lobby local cities to ease their permitting requirements to speed up and lower the cost of such housing. Several suburban cities pledged to do so, but Seattle was noticeably silent.

Homeless housing and low- to mid-range housing are related but meet different needs. Both are needed, but in my view, homeless housing meets a more urgent need. The costs of a worker housing program that would make a significant difference also seems financially much more daunting. Microsoft's priority on low-income rather than homeless housing may be a reflection of its suburban location—although it is also a serious contributor to efforts to ameliorate homelessness.

In the spring of 2021, a group of community leaders moved forward to place an initiative on the city ballot for November which attempted to balance providing mental and drug services and housing for homeless while removing tents. Even if adopted, its success would depend on implementation and funding by the mayor and city council, which was by no means assured.[19]

At the same time, Challenge Seattle began a renewed examination of homelessness by commissioning a new study by BCG which focused on solving the problems of chronic homeless people. BCG estimates this group to be about 3,355 people in Seattle/King County and growing at the rate of about 8% annually over the past four years. Chronic homeless are people with serious psychiatric/emotional conditions (73 percent) or substance abuse problems (64 percent) and have been homeless for extended periods of time. BCG's recommendations include emergency housing and on demand individualized services. They point out similar successful programs in San Diego and Bakersfield.

Several giant tech companies and their founders are stepping up to provide funding for housing for homeless and worker housing. In addition to Microsoft's $450 million loan guarantee program, Google has announced a billion-dollar program of low-interest loans, guarantees, and grants. Bezos is donating $1 billion to nonprofits across the country providing services to the homeless, and Amazon has announced a $2 billion low-interest loan fund for "very low" to low-income housing.[20]

Amazon also has built a two-hundred-room shelter for homeless families in one of its downtown office buildings that is run by the nonprofit Mary's Place. Amazon also provides twenty-five thousand square feet for a full-service restaurant, three cafes, a coffee shop, a catering venue, and training classrooms for FareStart, which trains people struggling to overcome poverty, addiction, homelessness, or criminal records for food service jobs. Paul Allen contributed $30 million toward another complex, operated by three nonprofits, that will house ninety-four families—half homeless and half low-income.[21]

Numerous nonprofits in the Seattle area also provide housing and services for the homeless. Notably, Plymouth Housing has constructed and operates fourteen buildings housing over one thousand residents.[22] Treehouse provides services to foster children, many of whom become homeless. Home Base, launched by the Seattle Mariners, has raised several million dollars from local companies, foundations, and individuals to provide emergency rental payments for a month or two to prevent evictions that would otherwise occur.

Notwithstanding this support from businesses and nonprofits, if Seattle is to raise the funds necessary to end homelessness, major global foundations headquartered in Seattle will need to join with major businesses to make substantially greater contributions.

A contributing factor to increased homelessness is the increased rents caused by the increase in workers attracted to work at growing tech companies which can price low-income people out of housing and into homelessness. Ordinarily, the demand for housing would be met by the construction of new housing. In cities such as San Francisco and Seattle, however, restrictive zoning laws, rent controls, not-in-my-backyard (NIMBY) resistance, and lengthy and expensive permitting and environmental processes unnaturally restrict and raise the costs of the construction of new housing.[23]

Seattle housing prices, although growing rapidly, are still substantially less than San Francisco's, where neighborhood groups have even more vehemently opposed new construction.[24] A few years ago, Seattle entered into a "grand bargain" with commercial developers, requiring them to pay fees toward affordable housing on any new project. New apartment buildings had to include units set aside for lower-income renters or else fund low-income housing elsewhere. In exchange, the city rezoned its commercial areas to permit taller buildings.

Seattle is beginning to rezone Seattle's single-family residential neighborhoods to allow apartments and backyard cottages. At the same time, however, they are consciously avoiding any consideration of rezoning an area of about five square miles in SODO (south of downtown) to allow high-rise apartments and office buildings. SODO is immediately adjacent to downtown with excellent access via walking, bicycles, and transit, including Seattle's light rail, which runs its length. The area, currently zoned for manufacturing and warehousing, holds Seattle's sports stadiums, parking lots, and a scattering of one- and two-story buildings.

Chamber of Commerce leaders, labor unions, and politicians have long protected SODO to preserve manufacturing and shipping jobs.

Seattle, however, long ago became too expensive for manufacturing companies, which are moving outside city limits—in many cases, out of state. In their place, SODO could easily provide mixed-use housing with incentive zoning to require allocations to homeless and low- and middle-income housing. At a meeting with Mayor Durkan, in late 2019 to present a proposal by Greg Gottesman and me to use Opportunity Zone financing for homeless housing, I raised the possibility of building on underused city- and transit-owned properties in SODO. Her quick answer was "not now" which I took to mean that she thought it would be too big a fight, even though it could help solve one of the city's biggest problems. Seattle is running out of space for software firms too; what if SODO could become home to Seattle's next Amazon?

Even though technology alone cannot solve homelessness, it can help. Over twenty years ago, before the consumer use of email, text messaging, and cell phones, voicemail was important. I was on the board of a Seattle public company, Active Voice Corp., that was a leader in developing and selling voicemail systems to businesses. Its founder and CEO, Bard Richmond, conceived and launched a free voicemail system for homeless people. This was important for them in arranging job interviews and obtaining social services because, without a voicemail box, people had no way to connect with them or leave a message. This program continues to this day and has served thousands of homeless.

Another example of the use of technology combined with an innovative business model is Housing Connector,[25] which has developed an online service to match homeless people with vacant apartments owned by small business landlords who often leave their apartments unoccupied when they are away for fear of damages, rent strikes, and other difficulties they associate with low-income renters. I recently met with their founders. Housing Connector provides a single point of contact for the owner in dealing with tenant problems and works with the owner and the tenant to resolve differences. They also guarantee payment of the last month's rent and up to $5,000 for tenant-caused damages to provide some financial assurance to the building owner. They are working with over fifty landlords and have filled over one thousand apartments, the equivalent of a $100 million public housing project. Zillow has agreed to integrate the rental program into its website to increase its availability to renters.

A major technology innovation, cross-laminate timber (CLT), can reduce the cost of building homeless and other housing by as much as 20 to 30 percent in place of steel and concrete. Forterra, a Seattle-based environmental nonprofit is planning to use it in new construction projects in Seattle (for Somali refugees) and Tacoma (for homeless). I discuss CLT in more detail in chapter 11 about new technologies for improving the environment.

PreK–12 Education and Beyond

Does it feel strange to pair a discussion about tackling homelessness with one about improving public education? It shouldn't. Both systems are mired in bureaucratic roadblocks that cause unnecessary suffering and prevent people from fully achieving their potential. In both, technologies and tech companies can help, but government leadership and innovation are even more important.

PreK–12 Public Education

Since the eighteenth century, America has embraced the belief that universal education is essential to the country's success. That philosophy is as true today as it was then. It is also the one means for addressing income inequality that all citizens can agree on—no matter their politics. This point was forcefully made by Rahm Emanuel, recently two-term mayor of Chicago and previously President Obama's three-year chief of staff, in his new book.

> Education reform . . . is the key to putting us on the path to solving what is perhaps the most pressing issue in the United States and maybe even the world today. . . . That issue is income inequality.[26]

Yet in Seattle, which has one of the best-educated and most progressive adult populations in the country, it is somehow acceptable that 62 percent of Black students in Seattle Public Schools are not proficient in reading by the sixth grade. For white sixth graders, it's 17 percent. In math, the disparities are even worse. As they finish middle school, an astounding 70 percent of Black students cannot compute at grade level, compared with 28 percent of white students.[27]

The school district is not unaware of the achievement gap, but their multiple plans over multiple years have done little to close the gap. Most recently, the superintendent proposed focusing on teaching reading to fourth-grade boys. An admirable goal—but she subsequently resigned because of the overall lack of support from the elected school board.

These deficits have relegated thousands of young people to a lifetime of low-wage work and have virtually guaranteed their ineligibility for the kinds of tech-based jobs that allow Seattleites a decent standard of living. We are missing an opportunity for raising the future income of tens of thousands of local families.

Amazon recently announced that it had ten thousand unfilled positions in the Seattle area. But few children of color here will graduate with the skills to qualify for them. The situation is much the same in other technology hubs like San Francisco, Boston, and New York City.

The waste in potential talent is staggering. Amazon is not unique; Microsoft and many other companies are eager to hire local kids, but so many are unqualified that these firms solve their needs by importing talent from India, China, Eastern Europe, and South America. In Silicon Valley, 60 percent of science and engineering workers with a bachelor's degree were born in other countries.[28] The CEOs of Alphabet, Microsoft, Tesla, and Uber are all foreign-born, and at least twenty of our eighty Madrona companies are led by first- or second-generation immigrants.

The failure of preK–12 public education in Seattle and other tech cities to lift low-income students has several causes. Some believe that more money is needed. Increased funding for more librarians, counselors, and social workers might help, and higher pay for teachers might attract more highly skilled people to the profession. But without fundamental reforms such as reducing bureaucratic centralization, giving more authority to school principals, and changing union-imposed seniority and other rules, increased funding would likely make little difference.

Some argue that the most important way to improve Seattle Public Schools is to eliminate "structural racism" in the schools.[29] This is a school system whose teachers would almost entirely classify themselves as antiracist and a district whose mission statement "recognize[s] the impacts of institutional racism on student success and question any excuses for not making necessary changes." One sensible step that has been suggested would be to "invest in a more racially diverse workforce."[30] We also need to address the effects of poverty and the breakdown of family units.

Teachers' unions have done much over the years to increase teacher's pay. But they also continue to stand in the way of needed reforms at the expense of children—such as opposing rigorous evaluations of teacher performance based on how well their students meet achievement goals and, when teacher layoffs occur, they require that the youngest teachers be terminated first. That is not fundamentally a money problem—it is about an entrenched and powerful bureaucracy that continues to resist educational reform.

Union pressures often limit attempts to improve schools and teaching. For example, the Seattle School District under pressure from unions prohibits Teach for America (TFA) teachers from working in Seattle Public Schools. These problems are not only endemic to Seattle. London Breed, the mayor of San Francisco, who is Black, has been outspoken in her criticism of the city's teachers' union, for being more focused on renaming buildings to reflect an equity-sensitive stance than on actually helping minority students. "The achievement gap is growing," Breed said, "and there are kids who live in poverty who are falling behind."[31]

Can tech help make sure that all students receive an education that allows them to succeed in life? Despite my years of advocating for the greater use of technology in schools, my feeling is that tech is not a magic bullet. Nonetheless, education would significantly improve with greater use of data-driven tools providing individual student tracking, increased access to the internet, and remote access to quality teachers and specialized courses. For example, at the college level, "flipped classrooms"—where a teacher delivers a lesson remotely, by video, and uses classroom time for coaching and applications—have been successful.

COVID-19 gave us new insights into how to use remote learning effectively as an important part of education. On the negative side, we learned that because many homes and children lack computers and broadband connections, children of low-income families are unable adequately to participate in remote learning. This means that they miss not only the formal learning of remote classes but lack the benefits of the vast trove of information and videos of nature, history, and people that can incite and inform. While some children were able to watch the SpaceX launch of the first American astronauts on a U.S. rocket in ten years, thousands were deprived of this kind of excitement that can inspire a career because of the lack of internet access.

COVID-19's exposure of this gap in such access has stimulated a renewed effort to extend bandwidth into rural and low-income urban areas and to provide laptops to students. But we have a long way to go to make this universal.

Seattle public schools struggled to quickly pivot from traditional in-person lessons to the use of online platforms. The first response to the question of remote learning was for Seattle public schools to shut down entirely in the name of equity: Seattle Public Schools' superintendent said that if all students could not access the internet for learning at home, then none would.[32] That call was soon reversed, and public and private efforts were launched to provide laptops to all children. But the fracas underscored legitimate worries about student inequities and the system's ability to address them.

When teachers' unions opposed reopening schools for in-person learning in several major cities during COVID-19 they were criticized by many of their traditional supporters for not adequately weighing the negative impacts on children, particularly minority children, because health evidence indicated that the COVID-19 risk to young children was relatively minor.[33] At the same time, once again, private schools and public charter schools were much more creative in finding ways to reopen in-person learning. It was not until April 2021 that the Seattle teachers' union agreed that middle and high schools could open for in-person learning—for two half days per week.

The forced digital response to the COVID-19 pandemic has created hope that the more widespread use of computers and tablets by teachers, students, and parents and the adoption of a variety of educational apps will carry over after students are back in physical classrooms. Michael Chasen, who co-founded Blackboard, believes this has "sped the adoption of technology in education by easily five to 10 years. . . . You can't train hundreds of thousands of teachers and millions of students in online education and not expect there to be profound effects."[34] Others echo the same view. "This is a disruptive moment" for schools, said Robin Lake, director of the Center on Reinventing Public Education (CRPE), which is partly funded by the Gates Foundation. There are so many discoveries, realizations—so much innovation," she said. "Public education will never be the same. . . . We're never going back fully to the old ways."[35]

Notwithstanding this optimism, I fear that with the end of COVID-19, public schools will revert to their old methods and not adopt many of the ways educational institutions, from primary schools to universities, implemented remote learning. We should be drawing lessons from these innovations to identify what worked and what didn't. Unfortunately, the culture of many school districts is bureaucratic and risk adverse, reinforced by the frequent opposition of teachers' unions to change. This is particularly true in big city school districts, but hopefully there will be more willingness in many of the thousands of school districts across the country to innovate and try new methods. The benefits of online learning are likely to be widely adopted in public charter and private schools, widening the gap in learning.

~

Here are a few examples of experiments that were tried during the pandemic that districts should consider continuing and adopting. In addition to conducting remote classes and increasing use of videos available on the Internet for class presentations and discussions, teachers tried new software apps, many of which can survive the return to physical classrooms like using artificial intelligence to adjust lessons to each child's abilities, online homework help and tutoring, remote learning for suspended students (who are disproportionately minorities), developing and administering quizzes, and virtual field trips. There were also easy to use drawing tools to sketch, for example, the solar system and apps for teachers to choose topical news articles or short stories for class discussion, with different versions of the text depending on a student's reading level and providing quick feedback to teachers on each child's progress. There is renewed interest in long-discussed proposals to extend the school year by shortening summer breaks of 12 weeks to 4 to 6 weeks.[36]

The worldwide audience for Google Classroom, Google's free class assignment and grading app, increased to more that 150 million students and educators, up from 40 million. And Zoom says it has provided free services during the pandemic to more than 125,000 schools in 25 countries.[37]

Tulsa Public Schools' program, Tulsa Beyond, is expanding opportunities for learning outside the school building through internships, through apprenticeships, and concurrent learning with higher education and technical schools.[38]

Use of Zoom and other remote tools will give schools an increased ability to focus resources on students, minority and otherwise, who need special attention. Schools are finding that it is easier to use Zoom to arrange after-school tutoring or homework help that fit to a student's schedule that may depend on school transportation or work at home.

There are new opportunities to deal with mental health issues remotely and one-on-one check-ins between students and mental health professionals over Zoom. Teachers can more easily check in with students who don't show up for class—or with their parents. In fact, most students and parents could use periodic check-ins. Today attempts to schedule in-person sessions are often frustrated by parent work schedules and childcare. More parents can attend all-school evening parent meetings.

In El Paso, Texas, the district is installing a districtwide online daily check-in system where students sign in on their phones or devices to say how they're feeling, so educators can have an early warning when a student needs help. An online Parent Academy in Georgia's Clayton County Public Schools sounds promising.[39] It coaches parents on topics like constructive study routines and how to monitor student progress. They also offer remote translation services for families whose first language isn't English. We have long known that a child's home life affects their education. Let's work on transforming more parents into strong collaborators in their children's learning. Increasing parent involvement in their children's education is one of the most important things we can do. One of the ways to do this is to prioritize parent convenience in school contacts.

It will take time to figure out what apps and programs actually overcome the chronic problems of quality and equity in the education of our youth. As in private industry, some experiments will fail but schools should learn from failures and figure out which are most helpful.

It's possible that even more fundamental changes may occur. The disparity between failures of the public schools to deal with the pandemic compared to the successes of private, parochial and public charter schools, as well as resistance of public schools to adoption of new learning apps and methods, are likely to increase the movement of children into these alternative school systems. Nationally, nearly a third of parents say they are likely to choose virtual instruction indefinitely for their children, according to a February 2021 NPR poll.

A couple of years ago, a young Amazon employee told me that he was so angry at what his local school was doing to his child that he wanted to sue them. My flip answer was, "Go ahead." A better answer is, "Let's continue to support reform efforts for our public schools while simultaneously financing alternative and supplemental opportunities for educating our youth."

Even when systemic changes of public schools that close the gap between white and students of color are too difficult to achieve in the near term because they are wrapped up in political controversies over the role and power of teachers' unions, accusations of racism, broken families, and poverty, it is nonetheless worthwhile to continue to press for reforms in the schools and at the same time financially support supplemental programs that provide enhanced education for minority and low-income students.

In response to the protests over the police killing of George Floyd, tech companies and leaders have announced strong support for the Black community in eliminating racism and discrimination. Equally important is the need for them to increase financial support for long-term improvements in education for disadvantaged students through public schools or otherwise.

Business Support

Tech companies, founders, investors, and executives throughout the country have created and support many programs to improve the education of disadvantaged students. I describe below several Seattle programs because they are good representations of what tech leaders can undertake, and I know them best from our family's support and involvement. These programs may provide ideas for other cities.

Technology Education and Literacy in Schools (TEALS), significantly funded by Microsoft, organizes volunteers with a strong background in programming to team-teach computer science alongside classroom teachers in schools that otherwise likely couldn't provide a computer science course. Their purpose is not only to boost access to tech for students but also to advance teachers' tech knowledge and skills.

TEALS serves nearly ten thousand students at 455 high schools in the United States and Canada. Over 1,450 volunteers from nearly seven hundred companies teach the courses who must commit to teaching a one-hour course at least three days a week so that these courses can meet on a regular schedule.[40]

Rainier Scholars, founded over twenty years ago, has provided eight hundred carefully selected low-income students of color with accelerated learning, academic guidance, and leadership development beginning in sixth grade through college graduation. The model is demanding, with the explicit goal for each student of acceptance by a college or university.[41]

Trish Millines Dziko, a former Microsoft executive, founded the Technology Access Foundation (TAF) in 1996, which has created a science-focused and demanding school of mostly low-income children of color who are admitted without using test scores or other metrics.[42] Though she'd wanted to locate her program in one of Seattle's already existing minority-majority high schools, public school teachers opposed her because they were worried about the stability of their jobs. At a boisterous public meeting, the local PTA also objected. The very next day, a suburban school district twenty-five miles south of Seattle invited Dziko, who is Black, to open there.[43]

Fast forward fifteen years, the school now has 640 students from sixth through twelfth grades. Seventy-eight percent percent go on to attend college—rates vastly better than the statewide average. Twenty years after refusing Dziko, Seattle has invited her back to launch a similar program at an inner-city school.

TAF operates essentially as a charter school within the traditional system, able to preserve its autonomy on teacher hiring—who remain members of the teachers' union—and maintain its project-based curriculum. The reason? Largely the endorsement of parents in Federal Way who supported Dziko's focus on math, science, and high standards. But it is also due to Dziko's tough, persistent devotion to an idea. As in technology companies, vision matters; leadership matters.

Sea.citi.org, modeled on Sf.citi.org, was founded by Amazon, Vulcan, Facebook, Google, and other tech companies to help solve educational problems in Seattle.[44] They have adopted eight low-income Seattle public schools to help them raise student performance. During COVID-19, they recruited 140 tech volunteers to assist more than 750 school families, including numerous families that speak languages other than English, to help them transition to remote learning, including the setup of 8,200 computers donated by Amazon.

Code.org promotes the teaching of computer science in schools throughout the United States. Its programs train and support teachers of computer sciences and deliver online courses and coding projects. Founded by a former Microsoft executive, Hadi Partovi, Code.org reached tens of millions of students in 2019, including strong participation by women and underrepresented minorities.[45]

Another national program, Amazon Future Engineer (AFE) is aimed at increasing access to computer science education for hundreds of thousands of children and young adults from underserved and underrepresented communities. AFE includes sponsorship of onsite professional development in computer science of elementary school teachers in more than one thousand school districts, funding for intro to computer science and AP computer science classes, college scholarships, and paid internships at Amazon after the first year of college. During COVID-19, Amazon also provided free access to online computer science courses provided by Edhesive.

When COVID-19 forced the closure of the Pacific Science Center, its leadership quickly swung into action by launching virtual field trips for preK–6 students and Curiosity at Home for all ages, with streaming live science shows, hands-on STEM activities, and more, to supplement what schools were offering.

It's not only tech leaders who are concerned. John Nesholm, a leading Seattle architect, and his wife founded a program that pays for reading tutors to work with seventh graders. When I met with Nesholm at our local Starbucks, I expressed astonishment: "Are you telling me that there are seventh graders who can't read?" John replied, "Unfortunately, lots." Could this program be scaled up with additional funding to cover all seventh-grade classes with a significant portion of nonreaders? Even better, could Dziko's new school in the Seattle Public Schools teach children of color to read at the same level as white students? If so, it would be a powerful model for reforming Seattle Public Schools from within.

It is also good to remember that opportunities also can be created by modest efforts by parents. Bill Gates and Paul Allen received their start because of donations by the Lakeside Mothers Club at a private school in Seattle that paid for a teletype machine that connected to remote computers.

These are all great programs. But unless teachers' unions can be persuaded to loosen some of their resistance to change, we will not be able to make material improvements to our public education system at scale. In the meantime, programs like TAF and Rainier Scholars must be expanded so that they reach a larger segment of disadvantaged kids. In Seattle, they represent our best chance for building a population of successful minority scholars, scientists, and businesspeople. Perhaps some will grow up to fix our public schools.

In many cities, there are often a few public schools—or programs within traditional schools—comprised of particularly talented students. New York has its famed Stuyvesant High School, Bronx High School of Science, and Brooklyn Technical School—which are now under attack as inequitable. Seattle, by contrast, has a highly controversial program for kids deemed "gifted." A *Seattle Times* analysis found that white children were eighteen times more likely than Blacks to be in classes for "gifted" youth with Black students making up less than 2 percent of Seattle's "Highly Capable Cohort," though they are 15 percent of the school district.[46] These inequities prompted the superintendent to announce a phase-out of the gifted program—again taking the route of reducing the opportunity for all, in the name of fairness—rather than find ways to elevate more kids of color into demanding classes. This is an artificial choice, as programs like Dziko's demonstrate.

In major cities like Boston, New York, Chicago, and Cleveland, control of local public schools has been transferred from an elected school board to a board appointed by the mayor. Although not a panacea for improving our schools, some studies have shown that academic performance is better in such districts.[47] But reforms are often limited by teachers' unions that often can politically outmuscle reason-based and public support for reform.

To obtain a quality education, parents too often have to move to the suburbs or, if staying in the city, compete to have their children admitted to special public schools or advanced classes or enter a lottery for a public charter school. If they can afford the steep tuitions for private schools of as much as $40,000 annually, their children can compete for limited openings.

Public Charter Schools

Attendance at public charter schools is free because they get their base funding from the state government. Businesses, foundations, and individuals often supplement their funding. Charters are less subject to the kinds of regulations that bind traditional public schools. They are run by an empowered principal reporting to an independent board. Charters are usually nonunion, and their principals have the freedom to hire, fire, and promote teachers on merit. Some are nonprofits, and others are run by corporations, several of which own and operate multiple schools. They may not discriminate in admissions, which are frequently by lottery because the schools are oversubscribed. A large portion of charter students are children of color and a majority of charter students are from low-income families who receive free or reduced-price lunch. Charters seek to combine the best of public and private school systems.

The number of public charter schools varies widely by state and city because state laws often limit the number; only twenty-two states do not impose limits.[48] In some states, such as Washington, the number is severely restricted, and under pressure from teacher unions, further expansion is often stymied.[49] Nonetheless, charter schools serve more than 3 million students nationally.

In more than 58 cities, charter schools represent more than 20 percent of all public school enrollment. In seven cities, charter schools enroll more than 40 percent of students, and in three cities, charter schools enroll the majority of students (New Orleans, Detroit, and Flint, Michigan).[50]

In Washington State, there are only nine charter schools, six of them in the Seattle area. The Bay Area, by contrast, has over 170 charters. Washington, DC, has 123 charters that educate about half of the city's ninety-eight thousand students in public schools. New York City has 267 public charter schools with 138,000 students and thousands on waitlists.[51] All of the New Orleans public schools are charter schools.

Opposition to charter schools partly depends on teacher's unions' strong opposition, although many progressives oppose charters based on their belief that they draw funding away from traditional public schools and weaken the system overall. The sharp division between those who love public charter schools and those who hate them often appears to be between tech and nontech people. Each is vociferous in their defense or attacks, which provide little room for agreement on whether they produce better outcomes.

Even though there are examples of poorly performing charter schools, there is much evidence that they often produce better education for their students, including minority and low-income students. In a study of elementary schools in DC, the top five were all charter schools as were thirteen of the top twenty-five. Waiting lists for the DC public charter schools exceed thirteen thousand students.[52]

Students who attend charters and their parents are strong supporters of their charter school, and minority parents and public officials are often their most outspoken supporters. Former President Obama and many Black leaders are strong supporters.

Early Learning

In recent years, there has been growing recognition that enrolling children in pre-K programs, substantially increases the ability of children to meet achievement goals in K–12 schools and beyond, as well as achieve higher-level jobs.

Many states and cities have begun to provide funding for increasing the availability of pre-K education. Based on neuroscience research on early learning at the University of Washington, Jeff Bezos' parents, Jackie and Mike Bezos, helped launch a successful campaign to approve a tax levy funding early learning programs in the county that includes Seattle. More recently, Jenny Durkan, Seattle's mayor, pushed through a ballot measure approving a property tax increase to provide pre-K classes for all Seattle students as well as two years of tuition at local community colleges and other post-K–12 programs.

Jeff Bezos separately announced a personal commitment of $1 billion to build and operate free Montessori-inspired pre-K schools in communities across the country. Bezos hired not an educator to run his schools but Michael George, a twenty-year veteran at Amazon and former vice president of Alexa and Kindle who brings management and innovation skills.

George Kaiser, the philanthropist I met with in Tulsa, is one of the nation's top funders and advocates for the lifetime benefits of early childhood education—because of research showing its beneficial impact on each child's student achievement and future earning potential. Early learning centers in Tulsa, funded by his foundation, provide year-round high-quality care and education to two thousand children in low-income neighborhoods. He told me, "Kids are not responsible for their circumstances, but their success is too often determined by circumstances."

Beyond K–12

It has become clear to everyone that our modern economy demands a level of education beyond the twelfth grade. But this does not necessarily mean that every person must go to college. In most European countries, vocational training is embedded in secondary and postsecondary school systems, offering a significant number of students the opportunity for skills-based training in the trades. In the United States, unfortunately, the college-prep focus of the past twenty years relegated vocational training to the margins, diminishing a legitimate and potentially enriching career pathway for thousands of young people. Robust vo-tech programs with apprenticeships still exist in rural communities, but they are anemic in most city school districts—despite increasing recognition of their value.

Brad Smith, president of Microsoft, has been a strong advocate of the Washington State Opportunity Scholarships that provide scholarships in Washington State's thirty-four community and technical colleges for associate degrees or certificates in high-demand trade, STEM, and healthcare occupations. In a separate program, it provides scholarships that can be used for eligible STEM or health care bachelor's degrees at public or private colleges or universities in Washington. In June 2020, Microsoft launched an expanded global skills initiative aimed at bringing more digital skills to twenty-five million people worldwide by the end of the year. By the end of the year, Microsoft had reached thirty million people and announced an expanded offering of programs.[53]

As mentioned in a prior chapter, Amazon offers to pay for all of its fulfillment workers 95 percent of tuition at community colleges for courses in nursing, IT services, and other high-demand jobs. It assumes that many of these workers will use education to launch careers outside Amazon.

University Education

It will be interesting to see if the need for universities to implement remote and mixed physical and remote learning during COVID-19 will result in any fundamental changes in how their education is delivered to students. Companies such as Coursera, which provide quality online courses by leading faculty members to all areas of the world, seemingly have not dented university attendance.

A leading example of remote university learning is Georgia Tech's master's degree of computer science, taught remotely through internet courses, for $7,000, about one-sixth the cost of the degree by attending courses at the university. About two thousand students enter the program each year.

There are tech-sector visionaries who believe that the four-year structure of courses and campus life is an anachronism. Craig Mundie, senior advisor to the CEO at Microsoft, has proposed that universities shift from four consecutive years of coursework to two years of general study, including the benefits of learning and living with other students, followed by a year of employment, and then an additional year or more of specialized classes in the profession the student wants to pursue.

Peter Thiel, the cofounder of PayPal and an original investor in Facebook, believes that universities are overpriced relics that impede innovation while saddling young people with debt that merely pushes them into uninteresting jobs to pay it off. Instead, he pays ambitious and entrepreneurial students to leave school and launch ventures of revolutionary "innovation that will benefit society."[54] Thiel's most prominent star is Austin Russel, who received a $100,000 fellowship in 2013, which he used to found Luminar that developed a low-cost lidar chip. Now twenty-five, his company went public in the fall of 2020 and is worth $3.4 billion. Not surprisingly, Thiel received six thousand applications last year from students at top schools like Harvard, MIT, and Stanford. There is of course precedent, for Steve

Jobs and Bill Gates dropped out of college—without the benefit of Thiel's $100,000.

I am not quite so radical. Universities need reform in terms of more use of remote learning by the most talented professors and reductions in the power of the university bureaucracy while preserving the best of in-person teaching, faculty/student interactions, and research. But I understand Thiel's frustration. Several years ago, I was appointed by the University of Washington's president to a special committee to recommend changes. Our even modest recommendations were fought by several administrators and came to naught.

11

Transportation and Environment

Metropolitan Transportation

Despite an image and ethos of environmentalism, leading tech cities are among the most traffic-congested in America. Boston, San Francisco, and New York all rank in the top ten. Seattle is not far behind. Pre-pandemic, commuters in these cities spent anywhere from fifty-five to ninety-one hours per year stuck in traffic during peak travel time, according to a study by INRIX.[1] If remote working in the future brings down the number of daily commuters, we should see an improvement, at least temporarily. The surge in downtown tech and office workers is partly to blame, though many programmers and engineers prefer to live close to urban centers where their companies are located, and more than half of Amazonians in Seattle get to work using public transportation, biking, or walking. Amazon provides them incentives to do so.

Traffic isn't a problem confined to tech-tropolises, of course. Most of the major cities in the United States are clogged with slow-moving cars and trucks that pollute our air, waste our time, injure and kill riders and pedestrians, and are a substantial cost burden on low-income workers and the elderly. Technologies, available now and soon to arrive, including electric and autonomous vehicles, can substantially help alleviate these problems. They are also giving rise to new business models, such as shared vehicles of Uber and Lyft that are already creating new transportation options for people.

Our primary transportation solutions have been building more roads and expanding transit systems. We have reached our limit in building roads in metro areas, and although light rail systems provide efficient solutions for millions of users, they are not effective in serving much of the population who do not live and work near a transit station. Although 23.1 percent of commuters within the Seattle city limits travel by transit (mostly bus), only 10.7 percent in the greater Seattle metropolitan area do so.[2]

Lower-wage workers, in particular, are often poorly served by transit, especially those who work after-hours when bus schedules are often spotty. These workers need timely, point-to-point service that doesn't require multiple transfers. Serving this long tail of users with transit is extremely costly, with estimates in the Seattle metro area being over $20 per passenger per ride. Substantially less could be used to subsidize such services by Uber and Lyft. If instead, these workers rely on their own autos, many are one accident from being without essential transportation.

To address these disparate transportation needs, cities, states, and transportation agencies need to adopt new long-range transportation plans that fully take into account new technologies and business models that integrate autos, public transit, shared vehicles, corporate vans, and buses, bikes, scooters, and pedestrians. These plans should prepare for the arrival of autonomous autos and trucks and aerial delivery drones. In the meantime, private companies are not waiting, and without public spending and transportation bond issues, they are proceeding with new shared-vehicle models, corporate bus systems, and autonomous vehicles.

Autonomous, Connected, Electric, and Shared Vehicles

Innovations in artificial intelligence, mobile phones, wireless apps, electronic maps, video, Lidar, radar, and batteries are making possible *autonomous, connected, electric and shared* (ACES) vehicles and services. Together they can dramatically improve the way we get around metro areas, lowering the cost of transportation, eliminating most traffic accidents, and making possible zero traffic deaths, improving the environment, reducing congestion, recapturing wasted time, and increasing accessibility for everyone.

The name *ACES* was coined by Bryan Mistele, CEO of INRIX, a traffic mapping and data company.[3] He and I cochair a Seattle coalition of business and policy experts that works to improve our transportation system by accelerating the adoption of ACES technologies in the Seattle area.[4]

The arrival of autonomous vehicles has been slower than many predicted (including me), and some predict that we should not expect autonomous vehicles for many years because of the technology, political, and customer challenges. But in the meantime, we are already benefiting from new ride-sharing business models by Uber and other ride-on-demand companies and the accelerated production of electric vehicles. Autonomous features being built into current automobile models are reducing accidents and saving lives.[5]

Rides on Demand

Millions of people are using car services provided by Uber, Lyft, and others, which are improving convenience, bringing down the cost of transportation, and increasing availability. Rides on demand represent new business models made possible by new technologies, including mobile phones, mobile payments, GPS, and digital mapping. The arrival of autonomous driving will greatly increase the benefits of ride services by lowering costs and reducing congestion through efficiencies and eliminating traffic accidents and their consequences.

Some cities and local transit authorities have welcomed ride services as an improvement on the archaic taxi and private car services and are working with them to supplement transit systems.[6] Other cities and many transit authorities have viewed them more as an enemy than a partner. For years, you could visit Austin and Vancouver and discover that the cities had banned these services. Not until 2020 did the Texas Legislature overturn Austin's ban.

Other cities are imposing special requirements on ride service companies that result in significant increases in the cost of rides for customers. Notwithstanding an image that ride services primarily benefit well-paid techies, rideshares significantly improve service to low-income workers, blind and other handicapped users, elderly, the infirm and children, as well as users not located near a transit station who are all ill-served by public transit and taxis. According to research collected during the pandemic, nearly half of all rideshares started or ended in low-income communities, home to many essential workers. And seven out of ten rides originated or ended at a grocery store, pharmacy, or doctor's office, underscoring what a lifeline these car services can be. Also, rideshare companies provided hundreds of thousands of dollars of free ride credits so that frontline workers could get to work, disabled people to service centers, and assault victims to shelters when public health fears reduced the availability of public transit.[7] Car services also provide needed part-time jobs for people who choose to supplement their incomes working on their own schedule as well as full-time jobs.

Seattle has imposed minimum wage and employee benefit requirements on ride services under pressure from Teamsters and other unions. According to university research studies, drivers make either $9.73 per hour after expenses (the New School) or $23.00 (Cornell University)—accounted for by different definitions of working time and assumptions on expenses.[8]

Cities and states are also imposing special taxes and fees on rideshares that increase the price of trips. The state legislature in New York has levied a tax of $2.50 on rides in Manhattan south of Ninety-Sixth Street, which will produce $1 million a day. Governor Cuomo's office justified their tax as "a positive step in our efforts to find a dedicated revenue stream for our subways and buses, as well as easing congestion in Manhattan's central business district."[9] Seattle's mayor, Durkan, has proposed a $0.51 tax per ride to be added to the existing tax of $0.24 per ride. The new tax will produce $25 million a year to help fund a new streetcar line and some affordable housing.[10] Durkan has also proposed a drop fee for delivery services.

To cover the increased costs imposed by the city, Uber has raised its rates in Seattle by about 25 percent, which will impose the greatest burden on low-income individuals who depend on the service.

Considerable criticism of Uber and Lyft is based on their impact on traffic congestion. One study found that rideshare vehicles accounted for a 180 percent increase in driving on city streets and that 60 percent of those passengers would have taken public transit, walked, or biked if the vehicles weren't available. Another study earlier found that rideshare companies accounted for more than half of the increase in San Francisco's swelling weekday traffic delays between 2010 and 2016.[11]

Ride service companies respond by pointing out that household car ownership and commercial trucking contribute a far larger amount to congestion. In San Francisco, 12.8 percent of total vehicle miles traveled were made in an Uber or Lyft; in Boston, 7.7 percent; in LA, 2.6 percent; and in Seattle, 1.9 percent.[12] They also point out that ride services replace what would have been the use of private automobiles, thereby reducing congestion spent trying to find parking spaces or taking roundtrips to airports to drop off or pick up a family member or friend. Many rides are shared by multiple customers, which lowers the number of cars on the road. As discussed below, the introduction of autonomous ride services has the potential to significantly reduce congestion.

Electric Vehicles

The replacement of gas-fueled vehicles by electrics, even with the growing support of public officials and the obvious benefit of cleaner air, has been slower than desired—primarily because of the lower cost of combustion engine vehicles and limited battery life of electrics. We also need a more extensive network of charging stations. Electric car technology is advancing quickly, and with increased production by major auto manufacturers, I anticipate that the cost of electric-powered autos will fall below gas engines within five years. When the lifetime cost of a vehicle is considered, the costs of electrics may already be lower. We are also beginning to see the impacts of commitments by major transportation companies like FedEx, UPS, and Amazon to replace their fleets with electrics. (Amazon has ordered one hundred thousand electric delivery vans from Rivian.) The impact of private action is vividly highlighted by GM's announcement that by 2035 all new autos and trucks will be electric.[13] Other automakers will follow, and their innovations and increased production will drive down the costs of electrics and lengthen battery lives. These business forces will prove to be more important than government mandates.

Autonomous Vehicles

The timetable for the perfection of new technologies is often uncertain and longer than predicted, but do not doubt that within ten years autonomous driving will be widespread on most highways and city streets, initially with a mixed system accommodating fully autonomous, partially autonomous, and human drivers. The implications for auto ownership, urban congestion, highway safety, and social equity will be profound and mainly positive.[14] On the negative side, autonomous vehicles will eliminate millions of jobs for drivers—trucking employs some 7.4 million people—nearly 6 percent of all full-time workers.[15] Cities fear that autonomous cars will reduce parking revenues.

Widespread testing and learning of AV use are underway with Tesla, Google, GM, Ford, and others. Amazon has stepped up its autonomous activities with pilots and large investments. Nearly every auto company announced plans to mass-produce autonomous vehicles by the early 2020s, but these predictions were significantly optimistic. When the technology will be ready for substantial autonomy is still not clear, and the timing of overcoming potential regulatory roadblocks is unpredictable. Even less certain is the adoption of new consumer behaviors such as car ownership being substantially replaced by on-demand autonomous car services. It took over ten years from the introduction of automobiles for horses and buggies to disappear in urban areas.

Reduce Accidents and Save Lives

Thirty-six thousand auto deaths occured in the United States in 2019.[16] It has been estimated that 90 percent are caused by drunk driving and other human errors such as driver distraction by cell phone use.[17] Many, if not most, of these could be eliminated by driverless cars. So it frustrates supporters of autonomous vehicles when political leaders promote the impossible goal of zero traffic deaths from conventional vehicles, while they simultaneously impede the advancement of vehicles that have a hope of achieving that goal.

Even today, you can drive autonomously on highways and major streets with your vehicle automatically staying in its lane, adjusting speeds, and braking with only your hand resting on the wheel. Notwithstanding sensationalist news coverage whenever one of the autonomous autos is in an accident, Tesla reports that its data show that using the partial autonomous features on its autos results in significantly fewer accidents than without.[18]

For 2010, the National Highway Traffic Safety Administration estimated that traffic accidents cost more than $200 billion.[19] It is surely more now with many more cars on the road. These not only are an inconvenience and a major cause of traffic delays but for low-income workers and families can be an unmanageable financial burden leading to loss of a job and homelessness.

Improve Equity for Low-Income Residents

The benefits of autonomous vehicle technologies, like shared rides, are not only for the wealthy but will improve transportation for everyone, regardless of age, disability, or income level. The cost of autonomous car services will be far less than the annual costs of purchasing, maintaining, operating, and insuring a private auto—and that doesn't count out-of-pocket charges for repair or accident-caused injuries.

Moreover, on-demand autonomous vehicle services to transit hubs and between home and work, shopping and doctors' offices, will greatly improve accessibility for the elderly, disabled, infirm, and workers not located within convenient walking distance of a transit system. It will give them independence, a reduction in social isolation, and access to services. Even when located near a transit stop, transit is often overloaded and has infrequent scheduling at night when many low-income workers are employed and who require one or more transfers. This is true for many who live outside the city centers but work downtown, and of course many places of employment are spread throughout a large metropolitan area whose outlying areas and small towns are poorly served by regional transit. We also may see low-cost autonomous van or minibus services—although many state laws presently prohibit such private services if they compete with public transit.

Reduce Traffic Delays and Congestion

Opponents of autonomous vehicles argue that they will lead to increased traffic because of use by youth, the elderly, the infirm, and others who currently are unable to drive. Additionally, if we make traveling in an auto cheaper, more pleasant, and safer, we may create more traffic problems and take ridership away from transit.

Even if congestion increased, however, it could well be more acceptable because the idle time could be used more productively and pleasantly. Supporters of AVs also argue that even if more people are traveling in autos, widespread adoption of autonomous driving will reduce congestion and traffic delays because the capacity of existing highways will be increased by closer fore and aft spacing between vehicles, more uniform speeds, and dampening of ripple effects. An additional lane or two on multilane highways can be added because of closer spacing. A few years ago, Craig Mundie and I published a report in 2017 proposing that a lane on Interstate 5, the main highway between Portland, Seattle, and Vancouver, be exclusively devoted to autonomous autos and trucks.[20] A study by MIT researchers showed that in a city like New York, a fleet of shared, autonomous vehicles would reduce the size of the current nonautonomous fleet by 30 percent.[21]

Increasing the efficiency of our highways would reduce the need for costly new roads and rail systems.

Recapture Lost Time

In reality, we will not know how congestion will be impacted until roads are full of autonomous vehicles, but my money is on less congestion—and even when congested, the driver can be using his or her time productively for activities other than driving—like dealing with work issues or for entertainment.

Traffic delays are estimated to cost drivers in Seattle $1,500 annually, an aggregate of $3.4 billion annually in unproductive time.[22] Autonomous vehicles reduce aggravation as well as the cost of congestion since vehicle occupants can turn travel time into productive time. The President's Council of Advisors on Science and Technology estimates this time to be worth $1.2 trillion per year.[23]

Improve the Environment

The widespread use of autonomous vehicles will provide important environmental benefits in fuel savings as well as allow improvements to our cityscapes as we recapture parking lots and on-street parking for public uses. Because the advent of autonomous vehicles will likely proceed alongside the adoption of electric vehicles, additional benefits of reduced air and water pollution will accrue.

Improve Public Transit

Even though public transit serves millions of commuters throughout the United States every day, not all residents live near transit stops nor can they reach their destinations without one or more transfers, resulting in many people forced to use their autos for single-passenger trips. Commuters who drive to park-and-ride facilities often find them full and must continue their journey in their single-passenger cars.

Transit supporters are concerned that the likely lower cost of autonomous transportation and convenience may mean that fewer people will use transit. But I'd argue that autonomous vehicles and ride services will increase the use of public transit by providing commuters with a low-cost, last-mile solution. While in Denver and a few other cities, public transit agencies have teamed up with mobile ride companies to offer such last-mile services, sometimes at discounted prices, many more need to overcome their reluctance to enter such partnerships.

Tech Company Van and Bus Services

While tech companies provide free or discounted passes to employees for public bus systems, they are increasingly supplementing public transit systems with their own bus and van services. These not only benefit employees but reduce single-passenger vehicles on our roads. In the Seattle area, Microsoft and Amazon have extensive bus and van operations and encourage bicycle commuting with investments in bicycle lanes and storage. Amazon has also eliminated garage charges to anyone—even nonemployees—who parks in its downtown garages on weekends or after hours to encourage off-rush-hour downtown shopping and supporting restaurants.

The advent of autonomous vehicles will spur still more innovative transportation solutions. For example, individuals may rent their autonomous vehicles to fleet operators of autonomous vehicle services—analogous to homeowners renting rooms through Airbnb. Also possible is that private ownership of vehicles will decline because of the availability and lower cost of autonomous vehicles from car services. At a minimum, many families will likely give up ownership of a second or third vehicle.

Drones

I am confident that aerial drone deliveries will become widespread within five years. They will decrease the number of vehicles on our streets, thereby helping ease congestion and reducing vehicle accidents. Drones will also cut delivery times and costs and reduce energy use.[24] Truck traffic represents about seven percent of urban traffic in U.S. cities and a congestion cost of $28 billion (17 percent of total congestion costs) in "wasted hours and gas."[25]

Drones are capable of fully autonomous operation using precise mapping, GPS, smart vision, and object avoidance. They can land in outdoor spaces adjacent to homes and on office and residential building rooftops. I can imagine that drones could even fly into a room through an open window; possibly, buildings will devote a room for deliveries (a drone port), which would replace many street-level trucks now delivering small packages.

Amazon has been developing its drones for at least five years in a building in South Seattle that has been reconfigured to create an indoor flight test space. Although its service has been held up by delays at the FAA in adopting regulations, the FAA in August 2020 issued rules creating a framework that allows Amazon and other companies like UPS and FedEx to launch drone delivery services.[26]

In other countries, government agencies have moved more quickly to allow at least small projects for drone delivery. In Australia, Alphabet's subsidiary, Wings, has been flying drones in an e-commerce test with deliveries of food to customers' homes in a suburban neighborhood.[27]

Additional uses of autonomous technologies will include marine passenger services such as on the bodies of water adjacent to Seattle, New York, and San Francisco. Whether autonomous flying passenger drones will become important in the next twenty years is uncertain, but I hesitate to say never.

Environment: Air, Climate, Parks, Streetscape

In the coming years, new technologies will significantly improve our environment—and ultimately reduce global warming as a risk. For example, many of the innovations in transportation will have significant environmental benefits by reducing air and water pollution, reducing energy consumption, and improving our cityscapes. Businesses and technology will make important contributions by converting their fleets of vehicles to electric, investing in the construction of solar arrays and wind turbines for electricity, using cross-laminate timber in place of steel and concrete in new buildings, and developing a new generation of small nuclear power plants.

The replacement of combustion engine–powered vehicles with electric vehicles will have an enormously beneficial impact by reducing air and water pollution. The use of autonomous vehicles will improve the fuel economy by more efficient driving, fewer delays stopping and starting, more uniform speeds, and lighter vehicles. Air pollution also will be reduced by the more efficient routing of traffic.

Drivers of autonomous vehicles will not be clogging streets as they look for parking spaces.[28] Instead, autonomous vehicles can drop passengers and pick up other passengers or, if no passengers are available, pull into what would be a mostly empty parking garage. That would also allow for the elimination of curbside parking in downtowns, adding more room for bus and bike lanes, or making sidewalks wider for pedestrians and restaurants.

In many cities, parking lots consume dozens of acres, which would become available for parks, housing, and other needed uses. It has been estimated that up to 30 percent of space in the central business districts of forty-one major cities is devoted to parking.[29] Autonomous vehicles will allow us to reconceive these cityscapes, creating more space for pedestrians and repurposing parking garages for rideshare delivery, and charging stations. This can hasten the transformation of our cityscape for parks, pedestrians, restaurants, and safe use of bikes and other micromobile services.

The use of cross-laminate timber in the construction of new buildings can reduce the costs of construction by as much as 15 to 50 percent in place of steel and concrete, depending on the type of building, and has significant environmental benefits compared to concrete.[30] CLT is made by gluing strips of wood together for wood beams and panels, producing a fire-resistant lower-cost building material and makes possible modular construction with tight tolerances off-site.

It is more energy-efficient than concrete and steel, with better thermal performance that requires less insulation for reduced heating and cooling costs. It is naturally produced, renewable, and environmentally beneficial because the trees store carbon while growing. It allows for modular construction off-site, resulting in less waste on job sites and fewer greenhouse gas emissions. Cement manufacturing produces up to 8 percent of man-made carbon dioxide emissions worldwide.[31]

I have invested as an angel, along with two other individual angels and Amazon's Climate Pledge Fund, in a new timber mill that will manufacture CLT in rural Darrington, Washington. Sponsored by the environmental nonprofit Forterra, the mill will sell up to half of its output at a 70 percent discount for low-income housing. That's an example of the ways businesses, venture capital, and nonprofits can support new technologies to address complicated social problems.

In the last few years, large tech companies have begun to demonstrate that with their technologies, management skills, and scale, they can provide significant help in solving local and global environmental problems. They are making significant investments to ameliorate climate change and global warming.[32] Even though governments have been sounding the alarm for many years and making pledges to take action, corporate action may prove to be more important. Businesses can often move faster with less encumbering bureaucracy. They are also more likely to take risks and invest in innovations while better managing their investments.[33]

"Bold steps by big companies will make a huge difference in the development of new technologies and industries to support a low carbon economy," said Christiana Figueres, former climate change chief at the UN and founding partner of Global Optimism, in response to an announcement from Amazon about its commitment to be net-zero carbon by 2040—a decade ahead of the Paris Accord's goal.[34] Long criticized as a laggard on environmental matters, Amazon has also pledged to run on 80 percent renewable energy by 2024, and recently accelerated to 100 percent by 2025.[35] Amazon is being joined in this Climate Pledge by other companies, including Microsoft which has an even more ambitious timetable for meeting the Paris goals.[36]

In partnership with the Nature Conservancy, Amazon launched the Right Now Climate Fund, committing $100 million to restore and protect forests, wetlands, and peatlands around the world,[37] as well as a $2 billion Climate Pledge Fund to invest in products and services that will "facilitate the transition to a zero-carbon economy."[38] Amazon also announced that it was ordering one hundred thousand electric delivery vehicles from Rivian, the largest order ever of electric delivery vehicles, with vans starting to deliver packages to customers in 2021. Amazon's order combined with a $440 million investment in Rivian and the purchase of autonomous vehicle company Zoox, shows one of the benefits of having large tech companies that can undertake meaningful initiatives.

Individuals like Gates and Bezos also have the resources, skills, and commitments to help solve many of these problems. In its twenty-year history, the $50 billion Gates Foundation has made enormous contributions on global health, environmental and other issues. Likewise, Bezos has begun to devote his energies and funds to helping solve some our most pressing problems, including the environment. In the last couple of years, he has established a $10 billion Earth Fund for combating global warming and pledged $1 billion each to homelessness and early childhood education. Gates is known as much for his most recent twenty years leading his foundation as his first twenty years creating and growing Microsoft. Expect the same with Bezos. Twenty years from now, he will be as well known for his second twenty years as his first.

Many politicians attack Big Tech companies for their bigness, but it is these large tech companies that can most easily drive transformational change. If every major firm pursued environmental programs similar to those at Amazon and Microsoft, we would surely see climate impacts as large as or larger than whatever may come through governmental regulation or spending.[39] Want to solve global warming? Look to innovation and private capital.

But the road is never smooth. For example, substantial public opposition, including from environmental groups, may continue to prevent nuclear power from expanding in the United States, even as advocates like Bill Gates point out that we will never be able to achieve sustainable zero carbon emissions from solar and wind alone. Gates has invested in an innovative approach for developing nuclear power plants.

There are plenty of social issues that government and businesses can tackle together, and I've outlined several in the previous pages—education and public safety chief among them. But public-sector inertia is a true hurdle. For that reason, I believe leaders in business and nonprofits must step forward to engage with these problems as never before.

Part 4

12

Government and Business
Conflicts and Cooperation

When I have asked tech CEOs in Seattle which issues most concern them, they always answered the same way: improve public schools and preserve our environment—including the urban environment of livable downtowns and neighborhoods. Recently they have added public safety, reducing homelessness, and a less antibusiness city council to the list. An ambitious list.

More mentally ill, addicted, and otherwise struggling people are on Seattle's streets than ever. In 2006, a one-night count found 7,582 homeless people who were "sheltered" and "unsheltered." By January 2020 (pre-Covid), that number had grown 49 percent to 11,751.[1]

At the same time, downtowns in cities like San Francisco, Boston, and Seattle have flourished for most people with the acceleration of their economic flywheels, driven primarily by the successes of tech companies. Talented young workers flocked to these city centers, attracted by restored and modern buildings for apartments and businesses, redeveloped waterfronts, new restaurants, bars, and entertainment and tech and related jobs. Creative people wanted to be close to each other and network. Even older people sold their suburban homes and moved downtown for convenient restaurants, arts, and shopping. But, unfortunately, homelessness and traffic

Figure 12.1
Combined flywheel *Source*: provided by the author.

congestion grew, and public safety worsened. Public education remained inadequate for their kids.

Now we are seeing signs of a shift of jobs and talented people out of city centers as they fail to solve their social problems and city governments stifle innovation and alienate businesses with regulations and targeted taxation. This was accelerated, as the COVID-19 crisis caused more people and businesses to consider remote working permanently.

I introduced the concept of the livability flywheel in chapter 8 as an important way to think about how to marshal the strengths of the private and public sectors to address the social problems of our cities. I also noted the linkages between the economic and livability flywheels, which can be visualized in the diagram (see figure 12.1).

As I discussed in chapter 2, the most dynamic way to think about the economic flywheels of Seattle and Silicon Valley is that creative people combined with innovation to beget startups. Those small startups—including not only Microsoft, Amazon, Google, and Facebook but also less famous ones—grew into successful companies that attracted talented people who generated new wealth. Many of those creative types, in turn, came up with innovations and launched new startups such as Netflix, Instagram, Redfin, and Rover—an ever-expanding constellation of companies. The flywheels kept spinning, faster and faster, such that both regions are now home to hundreds of technology startups—most with names you've never heard of—adding millions of dollars to the tax base, alongside their Amazons and Googles.

This is also happening in Boston, home to great universities and hospital systems, with important technology offices of Salesforce, IBM, and Amazon Robotics; New York City, which still has five thousand Amazon employees working there despite its ugly breakup with Amazon, alongside seven thousand who report to Google, bringing tech skills to the city's long-established finance, media, and advertising sectors; and Austin, which has attracted major facilities for Apple and Tesla.

Even though the economic flywheel has provided jobs and created wealth for spending and investment in leading tech cities, it has not solved the social challenges of homelessness, inadequate education, public safety, and transportation. These need addressing by the livability flywheel, targeting social ills and using social innovation as well as economic drivers.

The livability flywheel depends on leadership from elected officials, corporate and nonprofit executives, and the community. They can lead the restoration of their downtowns with parks, housing, arts, schools, and jobs. This will attract talented people who create and staff businesses and provide civic leadership. All of which strengthen the livability flywheel.

The two flywheels have several common elements. Both depend on talent that produces innovation and leadership. Both depend on a dynamic, growing business community that produces wealth that finances livable cities and universities and schools that produce talent. And both depend on a livable city that attracts talent and businesses.

Cities need both flywheels to thrive. Although Seattle and San Francisco were highly successful in using economic flywheels to develop their modern tech economies, they are now struggling to reverse the slowing of their livability flywheels.

The economic and social devastation of many large and small cities in the later part of the twentieth century, from housing segregation, jobs fleeing to Asia and the southern United States, inadequate public schools, and the auto-driven migration to the suburbs, destroyed downtowns and reduced tax revenues. This slowed the establishment of flywheels and their initial spin. But more recently we have seen progress in many cities across the United States—even in the former Rust Belt of the Midwest. As I related in chapter 7, smaller cities you might not think of—like Tulsa, Oklahoma City, and Kansas City—are getting their flywheels spinning.

There are exceptions, however, like Detroit and Hartford, where progress is painfully slow or nonexistent. At its peak in 1950, Detroit was one of America's most populous cities, with 1.8 million people compared to today with 670,000 people, a median household income of $29,481, and 275 murders in 2019, the highest level of violent crime in the country.[2] This is a deep hole to climb out of, but there are early hopeful efforts by local leaders.

Notwithstanding the challenges, any city can create an economic flywheel tailored to its particular strengths. The most successful will thoughtfully leverage their advantages—be they location, livability, natural resources, an anchor industry, or talent—while tapping creative thinkers to invent and start companies.

Whatever city you're in, support for startups will be critical—they produce many of the ideas, entrepreneurs, and inventors that will drive the flywheel. The lifeblood of startups is talented individuals, risk-takers, and early-stage venture capital and angel investors.

Even in cities that have built strong economic flywheels, it has not proven easy to sustain the livability flywheel, as friction develops from inabilities to solve problems of K–12 education, homelessness, and transportation. Often, the lack of effective local leadership, political, business, and civic, is a major contributor to the friction.

Additionally, many leading tech cities are pushing back against new technologies and business models on issues such as video surveillance, rides on-demand, and autonomous vehicles—notwithstanding that technology innovation and business models bring great promise in contributing to the solution of civic problems. While most major cities across the globe have installed video surveillance cameras to deter, find, and prosecute criminals, San Francisco and Portland have banned their use by government agencies and the police. Seattle has removed the few public video cameras that were installed by police—other than for traffic violations. Restrictions imposed by cities on operations of mobile car services such as Uber and Lyft include minimum wages and regulations of working conditions for drivers.[3] They are also imposing per-ride fees that raise the price of rides for customers and setting limits on the maximum percentage fee that GrubHub and other meal delivery services can charge, which also raises the cost to customers.

Tech cities such as Seattle and San Francisco have attracted wonderful talent from across the globe and have created enormous economic benefits, but they are endangering themselves as social conditions worsen and conflicts between the government and business increase. So, it bears asking:

Why can't cities and businesses that nurtured world-changing innovations work together to fix their social ills?

One barrier I have mentioned is the Seattle Way—our much-lampooned habit of discussing ideas to death without ever taking action. Or, put more bluntly, "the usual Seattle process of seeking consensus through exhaustion," as the *Seattle Weekly* described us back in 1983.[4] So extreme is the Seattle Way that it has its own Wikipedia entry: "the pervasively slow process of dialogue, deliberation, participation, and municipal introspection before making any decision." Recently, a major bridge in Seattle carrying thousands of autos and commercial traffic a day was ordered closed because of large cracks that had developed in the concrete. The bridge urgently needs to be replaced, and the mayor's response was to appoint a thirty-one-person task force to figure out what to do. Fortunately, I am not a member, but I am a member of the State Work Group on Autonomous Vehicles with seven subcommittees and a thirty-four-member executive committee. We meet quarterly in a cavernous conference room. As I mentioned back in chapter 10, another example is our region's homeless governing authority with a twelve-member governing committee and a thirteen-member implementation board, which one year after its formation had not hired an executive director nor made any significant decisions.

Soon after her election, Seattle's mayor appointed an Innovation Advisory Council of citizen and business leaders. A well-intentioned effort. This is an oft-tried-and-mostly-failed government response to unsolved problems, with its success thwarted by the lack of sustained leadership and the Seattle Way.

More serious is the growing conflict between major city governments and businesses. Escalating tensions dramatically came to a head in Seattle in 2018 after its city council proposed that our largest corporations pay a $500-per-worker "head tax" that would go toward housing homeless people. The tax was aimed at Amazon. In response, the company immediately halted several office construction projects already underway, threatening not to resume if the head tax was approved. Seattle was loath to think of itself as a company town, beholden to one big business. But this was no idle threat; Amazon had already announced that it was seeking a new city to host its second headquarters. Within days of the city council's head tax proposal, unionized construction workers, seeing their livelihoods in the balance, began protesting.

"No taxes on jobs!" they thundered at one councilwoman, a self-identified Socialist who had been hailed as a champion of workers. Seattle businesses—large and small, tech and nontech—joined in opposition to the head tax. Citizens and business groups launched a petition drive, collecting forty-five thousand signatures—twenty-eight thousand more than required—to get a repeal initiative on the ballot. Even the Service Employees International Union (SEIU) reported to the mayor that a majority of voters, despite their handwringing over the homelessness problem, were opposed to the tax.

The turnaround in a town long known as a blue-collar stronghold was not lost on anyone. The city council and mayor reversed themselves and dropped the tax. But the damage had been done, and Amazon began to consider capping its growth in Seattle.

"The city does not have a revenue problem—it has a spending efficiency problem," an Amazon spokesperson said, adding that Seattle's "hostile approach" to business was forcing the corporation to "question our growth here." A significant number of Seattle residents were sympathetic to that position. An independent poll in October 2019 before the election found that 64 percent of voters disapproved of the city government's handling of homelessness.[5]

Privately, Mayor Durkan expressed surprise at the business community's concerns—unwilling, or unable, to hear the concerns voiced by several CEOs at a meeting I'd attended only a few months before the 2018 tax attempt. They'd decried what felt like a virulently antibusiness attitude on the council. But Durkan waved off this criticism. Seattle was "not doing too bad a job compared to other big cities," she'd said.[6] Still, I couldn't help noticing that the mayor had been accompanied at this meeting by a sole junior staffer. And afterward, there was no follow-up.

This divide between government and businesses worsened in Seattle with city council elections in 2019 that pitted moderate, business-friendly progressives against extremists who won five of seven seats (including the reelection of the Socialist who'd incurred the wrath of construction workers). Soon afterward, as the city was convulsed by Black Lives Matter protests, this majority moved to cut the police department budget by 50 percent and lay off a hundred officers. That saber-rattling eventually resulted in more modest trims. But not before the Black woman police chief quit, unwilling to lead a police force under attack from city leaders even though the force was working to implement reforms.

COVID and Black Lives Matter highlighted many of our serious problems; however, our problems were rampant well before. For years, Seattle's city council had failed to deal with our thorniest social problems, rewarding crony social groups that supported them, echoing the crony capitalism they so often decried. When their consultants found that 30 percent of their sacred cow nonprofits had been ineffective in moving people from shelters to independent housing, the council voted to continue funding them anyway.[7]

Seattle then should not have been surprised that Amazon began to take actions to cap its growth in Seattle—beyond its previous announcement to establish a second HQ city and plans to open engineering offices of at least five thousand workers in each of ten offices spread across the United States, Canada, and abroad—originally driven by a desire to open up new pools of talent rather than politics. Most significantly, Amazon announced and is carrying out plans to locate twenty-five thousand software engineers and other employees to the nearby satellite city of Bellevue where the local government is welcoming.[8] Good news for the greater Seattle region, less so for Seattle. I don't think Amazon's new office diversification has to be a death knell for Seattle, even though I expect that most of Amazon's growth will occur outside Seattle's city limits.[9]

The Bay Area is facing these same questions with Google, Apple, and Facebook increasingly looking elsewhere for more fertile ground to grow their businesses. Many employees, no matter how well-paid, no longer favor working in an exorbitantly expensive, congested region that cannot solve its problems of homelessness and housing prices. With the new trends in remote working, downtowns are likely to have to compete based on livability.

For more than a few city leaders and voters, cooling off our Amazon fever would be a relief. A segment of Seattleites has long advocated a return to the slow growth of the past and an embrace of low tech. Of course, they also revel in the coffeehouses, restaurants, Uber rides, workout centers, bikes-to-go, and general big-city amenities made possible by the growth of a tech economy.

An important step in bridging the divide between the business and political communities is reaching a common understanding of the roles of government and business in solving urban problems. It has been long expected that the role of businesses was to produce jobs and pay their fair share of taxes to support the city, which was responsible for public safety, public schools, a safety net, zoning, public amenities such as parks, and transportation planning. Of course, it was never this tidy. Businesses bridled at what they thought were unfair taxes and often opposed government regulations and planning. Governments too often have failed in their tasks.

Cities have seldom been as idyllic as we sometimes would like to imagine. They were often slum-infested, with rampant discrimination, polluted air and water, and snarled traffic. Perhaps, to our credit, we are today less tolerant of problems of schools, homelessness, traffic, and public safety. And these problems have worsened in recent years. We are frustrated because, with our creative people, wealth, and technologies, we should be able to solve these problems.

In years past, we welcomed the prospect of Seattle's businesses leading the search for civic solutions. In 1968, a coalition of corporate leaders and engaged citizens launched successful voter initiatives to build new parks and pay for the cleanup of Lake Washington. In 1971, citizens launched an initiative that defeated a disastrous city council plan for demolishing Seattle's famed and historic Pike Place Public Market and replacing it with a complex of apartments, office buildings, hockey arena, and four-thousand-car parking garage. Today the public market is bustling, serving tourists and locals with dozens of vendors of produce, fish, cheese, meats, and flowers.

In the early 2000s, the Gates Foundation funded a $10 million transportation study by the Cascadia Center for Regional Development of the Discovery Institute offering solutions for replacing Seattle's aging Alaskan Way Viaduct, an elevated highway along the waterfront.[10] Everyone had recognized for years that the viaduct was an eyesore walling off the waterfront from downtown and an earthquake hazard; no one could agree on a fix.

Political leaders dithered and waved off the study's recommendation to rebuild the highway with a tunnel beneath our city streets. Flat-out crazy, local leaders and transportation experts said. But after the Cascadia Center organized conferences to highlight examples of tunnels in other parts of the world, flying in experts to answer questions, then-Governor Gregoire endorsed their recommendation.[11] The viaduct was demolished, and a two-mile tunnel funnels drivers below Seattle's downtown on a four-lane underground road that has opened up the waterfront to downtown residents, office workers, and tourists.

Another example: In the 1990s, Microsoft cofounder Paul Allen purchased a swath of underdeveloped downtown real estate, intending to create Seattle's version of Central Park. At the time, this was a blighted area of warehouses, abandoned buildings, and parking lots. But when voters twice rejected the necessary bond issues, Allen changed course, developed the property, and attracted Amazon to create an urban campus. Now fifty thousand workers have energized the neighborhood of restaurants, shops, and apartments.

So, we've done it before, leveraged private financial assets and leadership to design innovative solutions for the public good. Are our current problems of homelessness, schools, transportation, and housing so different from cleaning up pollution, redeveloping blighted streets, or solving the riddle of the viaduct? I don't think so. A hopeful sign is that technology companies and individuals are increasingly using their money, leadership, and technology to help solve civic problems, including homelessness, worker housing, PreK–12 education, transportation, and the environment. But it is a big, complicated agenda that needs increased support by government, foundations, and private businesses and leaders.

In 2014, a coalition of CEOs from fifteen of Seattle's leading companies and nonprofits created Challenge Seattle to address city and regional issues like education, homelessness, and transportation. Among its members are CEOs of Microsoft, Boeing, Nordstrom, Starbucks, Alaska Air, the Bill & Melinda Gates Foundation, and the Fred Hutchinson Research Institute. I have been a member from its founding, representing Madrona Venture and on behalf of smaller tech companies. Former Washington governor Christine Gregoire was hired as its CEO.

We hold quarterly dinner meetings, and a key founding principle was that only the CEO members may attend. This is different than most business civic organizations where it is common for attendance to be mostly by lower-level executives. This can work fine for those organizations, but for Challenge Seattle it has been highly useful to have a place where CEOs discuss issues directly with each other.

We have good attendance at our dinners, with at least ten to twelve CEOs attending. A CEO might discuss the concerns of their employees. Another CEO might discuss how they handled reopening in China and plans for the United States. Informal conversations occur over a glass of wine at the beginning of the meeting. Substantive reports and issues are discussed during and after dinner. Between meetings, corporate representatives work on specific programs.

When the COVID-19 crisis struck, Gregoire launched a twice-weekly one-hour video call of community and corporate leaders where she was the MC of sessions with the local health care and political leaders. These town-hall calls proved to be highly successful as accurate up-to-date sources of information on healthcare issues and resulted in recommendations to the governor for reopening schools and bringing workers back to their workplaces. Over 150 business, government, and health care leaders joined the calls. The CEOs also had joint calls among themselves on lessons being learned and plans for reopening offices. We also endorsed specific plans for eliminating discrimination in our workplaces.

Since its founding, Challenge Seattle has initiated and backed useful projects involving K–12 education, transportation, housing, and increasing crossborder trade with British Columbia, including a proposed regional plan for the corridor between Seattle and Vancouver. Going forward, post-COVID, I would like to see Challenge Seattle marshal the power of its CEOs to undertake additional programs that could bring about fundamental changes. My favorite example is education, where improvements would provide enormous, long-term benefits to children of color and our city. To its credit, Challenge Seattle launched a program in early 2021 to provide needed leadership on homelessness.

Challenge Seattle, however, needn't be our only vehicle for moving ahead to solve social problems. There are already dozens of corporate and nonprofit, as well as governmental programs aimed at solving problems. The challenge, seems to me, is how do we unite the necessary governmental, business, and citizen support to drive these programs to make a big difference?

Viable solutions will not arise from a watered-down consensus. What we need is leadership—ideally from a strong, politically astute mayor who can articulate the urgency of a problem, rally sufficient support, and organize governmental and private sector resources to execute solutions. Ideally, he or she would identify the community's most important problems, analyze the root causes, and implement innovative solutions.

~

But we can't always count on leadership from an inspired mayor. As an alternative, we can encourage and support individual and business initiatives. I believe the most effective approach is to create separate coalitions of businesses, nonprofits, and citizens focusing on specific problems as we have in the past with our history of meeting challenges like polluted water and urban blight. I'll be blunt: groups encompassing diverse and contradictory opinions have difficulty pulling on one rope in the same direction to get anything done. But single-purpose coalitions *can* gather people who are passionate and knowledgeable about particular problems, willing to devise new solutions, able to focus their fundraising, and willing to endure opposition. Empowering such coalitions to make timely decisions and take decisive action is much more likely to achieve positive results than time spent on consensus-driven task forces plagued by dug-in loyalties, skepticism toward new ideas, and obstructionist activists.

Although many projects can be driven by businesses with minimal help from the government, the best outcomes will come from businesses and governments working together. Leadership for such coalitions can often come from the business world, but it can equally come from passionate nonprofit leaders and citizens.

I hope we are past the issue of whether businesses should play activist supporting and leadership roles. Legal scholars love to debate the theoretical role of corporations in civic life, pondering whether their raison d'etre is solely profit, or whether their fortunes rise and fall with those of their home communities and other constituencies. The national Business Roundtable, composed of 181 CEOs, recently redefined the purpose of businesses as no longer only for the benefit of shareholders but also for "the benefit of all stakeholders—customers, employees, suppliers, communities and shareholders."[12]

There are business critics of these broader purposes,[13] but it seems obvious that modern businesses cannot be successful unless they treat their customers and employees well. Nor will they attract and retain creative thinkers in cities with mediocre schools, poor transportation, and rampant homelessness. If recent history is any guide, it is unlikely that these problems will be remedied unless businesses actively participate in the solutions.[14] The way I see it, cities need the risk-taking innovation and creative brainpower of businesses that are central to our tech economy.

This position is, of course, heresy to those who believe government alone is the proper body to decide on civic priorities.[15] In San Francisco, Marc Benioff, the CEO of Salesforce, argued that corporations, including his own, should do more to help solve urban problems.[16] His view was challenged by some on the left: "Relying on the privileged classes to set the social agenda during divisive times harkens back to the colonialism that the U.S. revolted against in 1776," warned Chiara Cordelli, a political science professor at the University of Chicago.[17] Similar criticisms have been leveled against the Gates Foundation for setting priorities for school reform and global health. But if the government has been unable to solve our problems, it seems logical, even imperative, to tap private-sector leadership and expertise.

Irrespective of these arguments, tech leaders are in fact diving into the civic arena with direct giving and action, along with a simultaneous distaste for funneling their money through government programs or generalized nonprofits. You can see this in Bezos's billion-dollar initiatives to target homelessness and early learning with projects that operate independently of government funding. His pledge of $1 billion to fight homelessness makes grants directly to nonprofits with a proven track record, and his money for preschools is to create a new nonprofit to build and operate them. Similarly, Microsoft's $425 million low-income housing plan is a corporate-organized initiative that does not depend on private-public partnerships—even though acceleration of building approvals would be helpful.

The response of businesses to the COVID-19 pandemic demonstrated how effective and important business involvement can be in helping solve major civic problems, particularly in supplementing and working together with government efforts. Critical to the development of COVID-19 vaccines and treatments in record time was the work of companies like Pfizer, Moderna, BionTech, Regeneron, and Eli Lilly in creating, testing, and manufacturing—all based on years of research at federal, university, nonprofit, and business labs. In response, the Federal Drug Administration showed that it could speed up its approval process while protecting public safety.

Although the federal government and most states faltered in their initial rollouts of testing and vaccines to sites for giving shots, they were essential to providing testing and vaccines on a large scale. Businesses stepped up and joined hospitals to increase the number of vaccine sites. For example, Amazon and Microsoft used their considerable logistical skills, software skills, and scale to quickly set up vaccination sites on their properties to vaccinate thousands of the general public per day.

Amazon also responded to COVID-19 by undertaking a broad range of other efforts to ameliorate its impact on people. Long-time Amazon critic, Scott Galloway, praised Bezos when he announced Amazon's $4 billion investment in COVID-19 products and services.

> Specifically, Bezos outlined a vision for at-home COVID-19 tests, plasma donors, PPE equipment, distancing, additional compensation, and protocols to adapt to a new world. Jeff Bezos is developing the earth's first "vaccinated" supply chain. . . . I believe Amazon will offer Prime members testing at a scale and efficiency that makes America feel like South Korea (competent).[18]

Amazon built its testing labs to be able to regularly test all its employees, including those showing no symptoms. They could expand this to cover people who are not Amazon employees and could grow into a general testing lab business, as well as at-home testing services. Amazon committed $20 million to its AWS Diagnostic Development Initiative, a program to fund diagnostic research projects that have "the potential to blunt future infectious disease outbreaks." Participating are thirty-five global research institutions, startups, and businesses.

Amazon began using its Alexa devices to help doctors diagnose diseases and allow users to ask health-related queries. Amazon joined the New York City COVID-19 Rapid Response Coalition to develop a conversational agent to enable at-risk and elderly New Yorkers to receive accurate, timely information about medical and other important needs.

> Our Alexa health team built an experience that lets U.S. customers check their risk level for COVID-19 at home. Customers can ask, 'Alexa, what do I do if I think I have COVID-19?' Alexa then asks a series of questions about the person's symptoms and possible exposure. Based on those responses, Alexa provides CDC-sourced guidance.[19]

This could be the basis for the future use of Alexa to provide further health information and even diagnostics.

Amazon Care, using Amazon's delivery network, teamed with the Seattle Coronavirus Assessment Network (SCAN), a University of Washington (UW) research project backed by the Gates Foundation, to deliver diagnostic test kits to Seattle residents who may have the virus for return through Amazon's drivers for analysis by SCAN and notice as to whether they are infected. This increased the availability of testing and allowed SCAN to study how the virus was spreading in Seattle. Amazon also collaborated with the Hutch on COVID-19 Watch, to trace the spread of the virus, with Amazon providing technical support, web hosting, and funding.[20]

In addition to their healthcare initiatives already underway, Google and Apple announced a joint project to develop an app for their cell-phone operating systems to track people carrying the virus, and the UW with help from Microsoft engineers further refined the app for use in Washington State. The developers specifically designed the app to protect people's privacy; nonetheless, states, local governments, and businesses in the United States hesitated for months in deploying the apps because of concerns of criticisms from privacy advocates. Many Asian and European countries, however, promoted and even required tracking apps which certainly saved many lives.

Asking individuals to give directly to focused nonprofits also can be powerful. My wife Judi and I participated with a small group of a dozen or so people who, led by technology leader Raj Singh and his wife Jill, raised over $27 million in one week in March 2020 for nonprofits directly helping people and small businesses harmed by COVID-19. We called our organization All In Seattle, but rather than having donors give the money to All In Seattle or to general community nonprofits to decide how it should be allocated, we asked donors to select and give directly to nonprofits who had active programs helping the needy.[21] The Seattle Foundation followed up with a broader-based program raising even more funds to help people being harmed by the economic impacts of COVID-19.

Though Seattle has a strong nonprofit community with dedicated leaders working on homelessness and public schools, their efforts need greater financial support from local businesses and foundations. More businesses need to follow the examples of Microsoft and Amazon. Many of our largest foundations, including the world's largest, the Gates Foundation, have a global focus. Even though they also are important contributors to local programs, I often think they should play an even bigger role locally with larger financial contributions considering their resources and that this is their home community from where they derived their original wealth and where their employees live and work.

The Kauffman and Kaiser Foundations that I talked with in Kansas City and Tulsa devote a substantial portion of their giving to local causes while still providing robust support for causes across the United States. As an example, Kaiser is restoring downtown buildings for restaurants, small shops, and businesses, building a business park adjacent to a historic Black residential neighborhood, creating early childhood education schools, and helping lead the community effort to develop the Gathering Place, an integrated sixty-six-acre riverside park with walking trails, attractive meeting buildings, and elaborate play structures. Cleveland is also well-known for its local foundations that support their community in a big way.

Is Seattle's dream dead? Does our inability to solve our city's problems threaten our ability to be able to continue to attract the creative people critical to our future? Or can Seattle's economic success be harnessed to support civic progress?

Can we have cities that are livable and welcoming for all, where people are not living—and dying—on our streets and where students from every economic bracket can rise to take advantage of the enormous opportunities that exist? Yes. But to do so, we need to make some changes.

Unless technology companies and cities find new ways to work together, the continued success of both is under threat. And they must smartly apply technology to help solve the problems. Just as businesses are using AI, cities should harness the ability of AI to analyze problems and propose and manage complex solutions. Cities should be leaders in making use of innovations in artificial intelligence, medicine, communications, and transportation technologies that are revolutionizing our lives.

Keeping our flywheels spinning is not a given. This moment is important. Business and political leaders must put aside their traditional antipathies. Politicians must dampen their antibusiness rhetoric, and business players must, as Bezos has stated, expect to be "scrutinized" by the press and government.[22]

Let's use Seattle's extraordinary prosperity and the growing creative classes in Seattle and San Francisco to attack the problems threatening to undermine our meteoric success. Let's institute a change in city government, opening its traditional processes to problem-solvers from outside. Then let's evaluate the results dispassionately and act decisively.

I don't believe any single entity—local governments, Amazon, the Gates Foundation, or anyone else—can solve Seattle's problems by themselves. But our old systems are not doing the job. And our culture of innovators seems an obvious, untapped source of new answers. We need this new generation of younger citizen, business, and government leaders who are not afraid of trying new solutions. We need to take advantage of our culture of innovation.

If we can't make tech cities great places to live as well as to work, we threaten the economic and livability flywheels built over the last generation.

13

The Future of Cities

It was the age of wisdom, it was the age of foolishness, it was the epoch of belief, it was the epoch of incredulity, it was the season of light, it was the season of darkness, it was the spring of hope, it was the winter of despair.

—CHARLES DICKENS, *A TALE OF TWO CITIES* (1859)

Charles Dickens could have been talking about today, a time of increasing turmoil and uncertainty even though paired with unparalleled technological opportunities.

Innovations in artificial intelligence, medicine, communications, and transportation technologies will revolutionize our lives. It will become routine to see drones flying deliveries through the air and landing near our homes or on office rooftops. Technology will supercharge crime detection with facial-recognition software that can comb through thousands of photographs in an instant, scanning for details as tiny as a small tattoo in search of a match.

Are there risks in the adoption of these new technologies? It's not hard to envision Kafkaesque scenarios of tangled databases, mistaken identities, and vigilante justice because of video services and giant databases. But the same technologies have the potential to increase the detection and prosecution of criminals and are already proving valuable in finding lost children and thwarting human trafficking. Artificial intelligence is capable of being engineered to eliminate racial and ethnic biases, making vision technologies more accurate than the fallible, distractible humans currently doing that work.

This fast-approaching world offers the promise of cleaner air (with more vehicles powered by electricity), fewer auto accidents (with autonomous vehicles eliminating human error), and more room for public parks and wider sidewalks for walking and outdoor seating (with fewer parking lots and on-street parking). Technology will improve healthcare through programs that analyze our medical histories and symptoms so accurately that we won't need to wait for a doctor's appointment to be diagnosed and to get treatment.

The innovations will continue to produce new wealth and make our lives better in many ways. But it is not inevitable that our urban centers will continue to grow with creative people, jobs, and wealth. The long history of cities is one of waxing and waning, from charming residential neighborhoods (and, yes, slums) and industrial and manufacturing strongholds to downtowns hollowed out by suburban office parks.[1] The modern tech industry established itself in the suburbs of Route 128, Silicon Valley, and Redmond, not downtown Boston, San Francisco, and Seattle.

Now, in addition to the reverberations of technology, we will be influenced by the COVID-induced experiences of remote living and working. Many companies, appreciating the benefits to their bottom lines and recognizing that it widens the available pool of potential employees, are making remote working an important part of their operations. Some are going 100 percent virtual without physical offices. Others are operating partly virtually, allowing many employees to work and live remotely, sometimes in cities distant from headquarters.

It's not that *everything* will change. People want and need to spend time together in person, mentoring, inventing, attending live concerts and ball games, visiting museums, and spending lots of time outdoors and in nature. There will be more e-commerce, take-out meals, and remote working, but there will still be a human need for the enjoyment of browsing physical stores, sharing restaurant meals with friends, and benefiting from the mentoring, socializing, and intellectual frisson that come through personal interactions at the office. Brick-and-mortar stores will remain as will physical offices, but there is little doubt that the adjustments wrought by COVID-19 will permanently alter our relationship to e-commerce, medical care, education, socializing, and work. Let's use technologies to free up more of our time for what matters.

Likewise, large cities are not going to disappear. Many people prefer to live in cities that offer strong cultural attractions, aquariums, zoos, restaurant choices, and diverse shopping experiences. Increased opportunities to socialize and network with friends and colleagues can be even more important. Businesses cannot completely abandon major tech cities because they must locate where talent wants to live. But there are increasing numbers of people who will choose to work and live remotely, in suburbs, smaller towns, and rural areas. This will not be insignificant to our future, but urbanologists are in the main correct—our long-term trend is toward more and bigger megacities spreading from downtowns through connected suburbs and towns. Think San Francisco/Oakland/San Jose/Palo Alto and Seattle/Bellevue/Kirkland/ Redmond/Tacoma/Everett.

The importance of building strong economic engines in cities and making urban areas livable will not diminish, and it will increase as metropolises face increased competition from other cities and towns that may be more livable and are building their economic flywheels. Importantly, major tech cities have lost their monopolies on talent for which they will have to compete.

The relative impact on smaller cities will be even greater as workers migrate to enjoy lower costs of living and fewer of the ills of the big cities. We are seeing signs of a shift of jobs and talented people again out of city centers as they fail to solve their social problems and city governments stifle innovation and alienate businesses with regulations and targeted taxation. Transplanted tech workers will provide a nucleus that over time will result in new startups being launched and wealth being accumulated for local reinvestment. Their presence will attract tech companies to open remote offices. The result? Their flywheels will accelerate.

The need for actions to improve opportunities for all classes of people was plain before COVID-19. Now it feels imperative. We have learned how to host business meetings, plan cultural events, and keep in touch with family via video telecommunication. Now we should incorporate these technologies into a better future for everyone.

To secure the benefits of remote work, education, and medicine, we need to expand broadband connectivity—especially to low-income communities and rural areas. We need to ensure that every public school provides each student access to a connected home computer.

We need to make healthcare research and investment a higher priority. Fortunately, we live in a time when scientists are cracking the genetic code, and over the next ten to twenty years we will be able to prevent and cure cancer and a wide range of other diseases. We should take advantage of the new attention by businesses such as Amazon and Microsoft and the power of AI to improve the delivery of healthcare and lower its cost.

More people working and learning remotely presents an opportunity to reduce traffic density in our cities. I have long been a proponent of autonomous electric vehicles, and I believe we should reconceive our transportation plans with better integrations of public transit, corporate buses, bikes, and auto-share services to prepare for the advent of driverless cars and trucks and aerial delivery drones. Instead of resisting this technology-driven future, local transportation agencies need to imagine redesigned cityscapes that eliminate on-street parking and convert open-air parking lots to parks.

In the United States, the pandemic exposed our unresolved conflicts between privacy and uses of technology as we failed for many months to implement tracing and tracking apps on our phones which healthcare experts have said are highly important to contain the spread of the virus, could save thousands of lives, and would help allow people to return to work and school. An Oxford University study found that if 15 percent of the population used the app, infections would be reduced by 8 percent and deaths by 6 percent. But as late as fall 2020, states were only beginning to approve and implement the apps.

The urgency of developing vaccines for COVID-19 showed us that we could shortcut regulatory requirements, overcome government and corporate bureaucratic delays, and increase cooperation between private industry and government. Nonetheless, in the future, we will still face continued disagreements about privacy policies, regulatory and bureaucratic delays, and conflicts between governments and businesses.

COVID also exposed how woefully unprepared our public schools were to meet the needs of students through online learning, particularly compared to public charter schools and private schools. The long-term ramifications of this gap in the education of millions of young people—felt worst among low-income children—are a profound concern.

The full impact of COVID-19 on remote work will not be evident for some years. But it is already clear that the forced experiences of working remotely will alter our workplaces and surely have an effect on where many workers choose to live. Not that I anticipate COVID-style isolation as our new normal. In-person conversations generate a particular kind of energy and serendipity that has value. But video communication has virtues too, by including a wider group in a meeting and in enormous time-savings by eliminating unnecessary travel even just across town for a meeting. As well, do we need to travel to every conference in a distant city? Businesses that don't need everyone in the office every day should modify their work processes so that workers can be remote at least one or two days a week. This would not only make them more efficient but free up our highways and public transit. Doctors who had resisted communicating with patients via emails or video calls tell me how they plan to use remote sessions as an important way they talk to patients.

The causes of the social problems of cities are complicated and disputed. Increased housing costs are the result of the demand created by economic growth and the people attracted by tech jobs as well as government policies restricting zoning and raising costs of construction. There is no one cause of homelessness, which results from drug addiction, mental health issues, poverty, and housing costs. Traffic congestion is mostly the result of economic growth but is often aggravated by local transportation policies.

Redlining has been illegal in Seattle for years and is morally repugnant to an overwhelming portion of the population. Its "Central Area" historically was mostly low-cost single-family houses for Blacks confined by redlining. As the city prospered, the convenient location of these houses to downtown caused middle-class and well-to-do whites to buy them and pour money into renovations or demolish them for more expensive houses, townhouses, and apartments. Low-income residents sold out and moved to lower-priced housing in south Seattle. Their children attend subpar local public schools where the school population is overwhelmingly nonwhite and poor. There is evidence that living in a low-income neighborhood makes it much more difficult to move up the economic scale than living in middle-income neighborhoods. Some cities have launched programs to subsidize housing for low-income residents in middle-income areas to try to address this imbalance, but subsidies will not be enough. Raising incomes through improved education everywhere will be more important and permanent for low-income and minority children.

Problems, such as ineffective education for disadvantaged students, which should be solvable, suffer from political barricades to reform from teachers' unions and politicians they have intimidated. The resulting disparity, as I have pointed out, is that 62 percent of Black students cannot do grade-level reading in the sixth grade compared with 17 percent of white students. There is evidence that the single best predictor of high school graduation is third-grade reading achievement.

The more immediate and visible problems of homelessness, discrimination, crime, and traffic seemingly distract us from solving the problem of low-quality schools. This was seen in polling results before the Seattle City Council elections in 2019 when voters were asked to identify two issues that they were most frustrated or concerned about. The results were: homelessness, 50 percent; housing costs, 27 percent; traffic, 22 percent; and crime, 10 percent. Near the bottom was schooling at 5 percent.[2] This from a city with one of the most well-educated white and Asian populations in the United States and appalling achievement scores of Black and low-income children.

As I've discussed throughout these pages, the development and growth of successful cities depend on two interdependent systems: the economic flywheel and the livability flywheel. The social welfare struggles in Seattle and San Francisco—and the beginnings of migration away from these cities—show that neither flywheel is sustainable without the other. As I hope the Midwest examples make clear, any city can create an economic flywheel by thoughtfully leveraging its advantages—be those locations, livability, natural resources, an anchor industry, or talent. The cities I visited have welcomed leadership from sectors beyond municipal government to get their economic *and* livability flywheels up and running.

Throughout the tech-led economic transformations of Seattle and San Francisco during the past twenty-five years, our government and business sectors operated on largely separate tracks, with minimal meddling in each other's affairs. That was how it should be, I always thought. But my beliefs have changed. When local governments prove unable to solve our civic riddles, it seems logical, even imperative, to tap private-sector strengths. We need leadership, as well as resources, from businesses, nonprofits, and citizens—to keep our flywheels spinning.

Some technologies have the power to change our lives irrespectively of whether city governments work to encourage or inhibit their use. We've traditionally looked to the government for transportation solutions of more roads and transit, but future solutions are coming from the private sector in new business models built on technologies such as Uber and Lyft and inventions such as autonomous vehicles. Private aerial drones will likely play a big role in deliveries. These innovations are not driven by government action, but governments can temporarily throttle them.

It is exceedingly difficult, however, if not impossible to create a livable city without effective actions by the local government. I have no special powers to predict the outcome of future local political battles and their impacts on the future of cities, but elections are important even if unpredictable and not always to our liking.[3] We need effective local political leaders willing and able to implement solutions to make our cities livable without inhibiting the growth of businesses and jobs. Mostly this depends on a strong mayor who, whether a Democrat or Republican, spends little if any time fighting the battles of the Congressional parties and pursues practical policies that solve local problems. Some of their policies are influenced by their philosophical bent, but often it is difficult to tell their policies apart—with Republican mayors welcoming immigrants and Democratic mayors disagreeing with teachers' unions. In his book, *The Nation City, Why Mayors Are Now Running the World*,[4] former Chicago mayor Rahm Emanuel highlights the similar policies of mayors of both parties in San Diego, Anaheim, Oklahoma City, and other cities. I would also cite Mayor Emanuel, himself, and Mayors Durkan of Seattle and London Breed of San Francisco, although all controversial, as the types of progressive mayors devoted to solving problems rather than sloganeering, who can be elected notwithstanding fractious local politics in their cities and even city councils that I would label as radical.

Unfortunately, not enough cities have a strong, enlightened mayor. Even with a weak mayor, it is still possible to get the flywheel moving with strong business and civic leadership investing in downtown and generating support for public improvements through citizen-led bond issues for parks and other amenities as has happened in the past in Seattle. Even with a forward-looking mayor, visionary business leadership is needed, since the investments in new businesses and attraction of a talented workforce come from business.

The radicalization of many city councils in leading cities has compounded their failures to solve pressing social problems. It is not a new occurrence that city councils tangle with their mayors and business communities. What is new is the often extremism of these councils. In Seattle, Portland, San Francisco, Los Angeles, Minneapolis, New York, Baltimore, and other cities, the radicalization is evidenced by votes to substantially defund their local police forces without effective alternatives.

When a majority of the Seattle City Council voted to ban the use of tear gas in limited circumstances, Seattle's police chief, who is a Black woman, pointed out that this denied the police nonlethal ways to control riots and all that was left to them was brute force and the baton. A majority of the city council followed by voting seven to two to reduce the police chief's salary in retaliation for her plainspoken defense of the need to have police to provide for public safety. Seven of the nine members also committed to defunding the police by 50 percent and as a first step required a reduction of up to one hundred officers, which under union seniority rules meant most of the terminations would be younger recruits, many of them Blacks hired to help reform the force. Ten million dollars of police funding was shifted by the council to community-based nonprofit organizations including activist groups that supported the council members' political goals.

The defunding of the police finally caused the chief of police to resign, say-ing she could not continue to lead a police force that was under attack from the leaders of the city council and harming the efforts of responsible people to reform the police while protecting public safety.[5]

How to explain what is happening in Seattle and similar cities? Not how to explain the protests and riots or the actions of the city council-members but why Seattle's well-educated voters elect them? And can this be reversed?

Maybe the best hope is that notwithstanding council divisiveness, effec-tive mayors can be elected citywide, as seen in the elections of Emanuel, Durkan, and Breed. In Seattle's 2017 nonpartisan election, Durkan, a liberal Democrat, defeated an articulate, extremist Democrat 57 percent to 43 per-cent, but at the same time, the more extreme candidates achieved a major-ity of the council. In the 2019 election, there was a hard-fought vote for council positions that pitted moderate progressives against more extreme progressives, one of who was a declared socialist, Kshama Sawant. In five of the seven districts, the more extreme candidates, including Sawant, were elected, giving the extremists a veto-proof majority of seven of the nine council positions. In a pre-election poll, only 37 percent of the voters approved of the job Sawant was doing. But four weeks later, she was elected with 51 percent of the vote. Perhaps most telling in the poll was that among eighteen to forty-nine-year-olds, more approved than disapproved.[6]

The antibusiness rhetoric of many members of these city councils is widespread, and Amazon cited this as one of the factors on whether they will continue to expand in Seattle. Amazon's decision to establish a second headquarters with twenty-five thousand employees in Queens was greeted with negative pronouncements by many New York City politi-cians directed at the economic incentives Amazon has been granted and more generally they directed a wide range of political criticisms against Amazon.[7] These included complaints that Amazon would not pledge to drop its opposition to union efforts to unionize Amazon employees and to selling facial recognition software to government agencies. The DC suburb of Crystal City across the Potomac in Northern Virginia was more welcoming for a second HQ.[8]

Even with the challenges of local politics, I am not predicting the end of cities as desirable places to live and work, maybe because I am an eternal optimist, but it will take determined civic and business leaders working to balance the extremists with strong mayors and, when possible, progres-sive councilpersons working together. Business and civic leaders must be willing to stand up to the extremists and advocate for programs that are effective and beneficial to the local citizens.

Government too often lacks the management skills and talented leaders to develop and implement solutions to critical problems and too often succumbs to political pressures. Business guru Jim Collins argued that "getting the right people on the bus" is the first step in generating momentum.[9] This applies not only to CEOs and business executives but also to mayors and their department heads.

The efficacy of every program—whether operated by the city government, nonprofits, or businesses—should be regularly evaluated and funding terminated for those that are persistently ineffective.

We need involvement and technologies from the business communities even in areas as uniquely a government function as providing public safety. In traditional government functions such as providing health, drug, and alcohol services, maybe in the longer-term private healthcare companies will perform these services better and at a lower cost. Nonetheless, successful local governments remain essential for many services such as police and fire protection, parks, fair taxation, and allocation of public revenues.

I am not ready to buy into the views of some such as Ray Kurzweil that AI will take over the world, eliminating the roles of political leaders and governments. Nor do I subscribe to the scenario in Yuval Noah Harari's bestseller, *Homo Deus* (2015), where software algorithms improve to the point that humans will choose to let them take over and make all of our decisions—personal, medical, and political. Both of those viewpoints prophesize a world where politics, cities, and our own choices will be ultimately overwhelmed by technology and rendered superfluous. Maybe that is the dream of some technologists. But the messy business of politics and human nature is a virtue. We must aggressively apply our human ingenuity to be the masters of our technologies.

Beyond the highly important task of helping create attractive, safe, and functioning places to live, there is very little governments can do to stimulate new business formation. Although some cities and states have successfully bid to induce businesses to locate offices and plants in their areas, such funds might be better invested locally in education to make their cities more attractive.

Yet without any help from the city or state, more than 140 global firms have opened software engineering offices in Seattle, employing some forty-five thousand people. We like to think it's because of our natural beauty or quality of life, but the most important reason is the depth of our talent pool—thanks to Microsoft, Amazon, and dozens of smaller tech firms.

My skepticism toward government solutions is not meant to minimize the value of having everyone contribute, from elected officials to civic leaders, corporate executives, unions, and workers. But public leaders must be held as accountable for results as any corporate leader. Let's open up the city government's deliberative processes to problem-solvers from outside. Let's evaluate the results dispassionately.

A bedrock goal of these burgeoning new economies must be social, educational, and job equities—without them, the livability flywheel slows, stumbles, and stalls. The need for intervention to improve opportunities for all people was clear long before COVID and Black Lives Matter. Now it feels imperative. Economic flywheels can be used to help achieve these ends. But to achieve equity in employment, it is not enough to have jobs for all. We must ensure that all people have an opportunity to get well-paying jobs. The key to that is education and, to my mind, it will require substantial reform in our public schools—starting with high-quality preschool for every child and ending with strong programs in our community colleges. These improvements will also improve college readiness for many more young people. Unfortunately, education reforms may take years, and often only minimal progress is being made in cities where it is needed most.

In the meantime, we can expand supplemental educational programs for minorities at all levels, as well as more robust use of remote learning and vo-tech programs and apprenticeship systems. Education and job-training experts must take the baton and look for ways to apply private-sector innovations in the application of adaptive software and data to tailor instruction to each student's strengths and use video and chat tools to integrate the best of in-person sessions with remote and digital learning. Teachers need to take advantage of connecting with people in faraway places, remote instruction by inspiring teachers, and integration of video segments of history and science that enhance learning.

The confluence of COVID, on top of increasing social problems in our superstar cities, shows me that the tech sector, for all of its success, does not exist in a vacuum. It is part of a symbiotic relationship with the metro areas where its workers reside. And increasingly, the disparity between haves and have-nots in these cities threatens the sustainability of our economic success. This is no longer an abstract concept. In Seattle, Portland, Minneapolis, Chicago, Philadelphia, Louisville, and many other cities, citizen anger exploded in the wake of George Floyd's murder by police. Seattle was convulsed in months of protest that often turned violent, with riots, fires, looting, and physical attacks on police.

From my perspective, these challenges, as well as opportunities, increase the pressure on our current tech superstar cities to get their livability flywheels moving, if only to better compete with the new generation of up-and-coming tech stars: Austin, Nashville, Miami, Denver, and cities like the Midwestern ones discussed in chapter 7. All of these places—with lower costs of living and less homelessness—are more livable for many people than San Francisco, Seattle, Boston, or New York, and they are already drawing talent away from those cities.

To build local economies and create solutions for our social problems, we need to make use of new technologies: AI, 5-G broadband connectivity, autonomous vehicles, and drones; intelligent applications; voice, motion, and touch access to the internet; robotics, quantum computing, virtual, augmented, and mixed reality; and biotechnology.

Notwithstanding our slow adoption of contact tracing and tracking apps, COVID prompted greater cooperation between government and private industry in areas like vaccine development. The urgency around accelerated drug design and approval shows that it is possible to eliminate a great deal of the bureaucratic delay that stymies forward motion. And while technology, misused, can infringe on personal privacy, the success of contact tracing and tracking apps in other countries shows that it is possible to balance society's needs with personal privacy.

We are too often stymied by mutual suspicion between businesses and government, dug-in loyalties to ineffective groups and bureaucracies, cautious government leaders, obstructionist activists, knee-jerk skepticism toward new ideas, and, at least in Seattle, by the action-killing deliberations of the Seattle Way. Unchecked, these barriers stymie even good intentions.

Tech companies, too, bear responsibility. They often have been reluctant to engage on civic issues. Fortunately, the leaders of many top firms have come to realize that they have an important and needed role to play in the lives of the cities where they operate. Whether that appreciation comes out of self-interest—the difficulty in recruiting talented employees to places with atrocious traffic congestion, out-of-reach housing, and subpar public schools—or whether it stems from altruism doesn't much matter in the end. Technology companies have tools that could help solve our transportation, environmental, housing, and public safety challenges, and they are beginning to put their money, leadership, and innovation skills toward those problems—rightly so.

If we are to achieve these goals, businesses need to step up and play a leadership role. It's happening in cities big and small. In Tulsa, the support of Oklahoma banker, oilman, and philanthropist George Kaiser, is funding incentives to bring new thinkers into that city, as well as early learning schools, parks, and urban redevelopment. There are lots of other examples. Microsoft has pledged to make low-interest loans to developers of $450 million toward building more affordable housing in the Seattle area; Amazon has followed with its pledge to make $2 billion of low-interest loans for low-income housing in multiple cities; Jeff Bezos unveiled a plan to invest $1 billion in creating subsidized preschool programs nationwide while giving another $1 billion to nonprofits fighting homelessness, and a group of tech-sector companies in Seattle has "adopted" eight of the city's neediest schools. MacKenzie Scott gave more than all of them—$6 billion in 2020 to over four hundred nonprofits in all fifty states like Black universities, food banks, and groups helping low-income children and those with disabilities.

These business leaders have realized that traffic, homelessness, poor public schools, and unaffordable housing—left unabated—will undermine their prospects for continued success. Their success also goes beyond dollar donations; they hinge on tech corporations contributing leadership and innovation.

Our cities' economic and social systems are interdependent. Government and local economies need flywheels that are aligned and each generating energy that benefits the other. Eventually, the failure of one will harm the other. That is why we need to build flywheels promoting economic and civic goals simultaneously. But when local governments are not sufficiently responsive and innovative, business and nonprofit leaders need to implement their own independent efforts.

The success of cities depends on developing and attracting talented people. Before the politics became unpalatable, Amazon was attracted to New York for its deep well of talent. Two hundred and thirty miles away, leaders in Crystal City, right outside Washington, DC, built their talent case by promising to invest $1.1 billion on expanding computer science education programs across Virginia, while simultaneously building a brand-new Virginia Tech graduate campus in nearby Alexandria.

Local talent of inventors and entrepreneurs are the source of new businesses driving startup economies. They come from local businesses and research institutions and often are immigrants from other parts of the United States and across the globe. Even advanced tech economies such as Seattle and the Bay Area depend on the continued attraction of talent. Generating home-grown talent depends on improving our K–12 schools and strengthening our universities.

Stanford University has been an integral part of the Silicon Valley technological hotbed since the mid-twentieth century when the region was better known for apricot orchards. Similarly, MIT is a major force behind Boston's inclusion on the list of top tech cities. The University of Washington nurtured a young Paul Allen who, after leaving Microsoft, went on to found five $100-million research institutes in Seattle focused on artificial intelligence and biology. Allen and Gates have also been major donors to the UW's Allen School of Computer Science & Engineering (CSE). Madrona has funded nineteen startups from the UW's CSE.

What about places that don't have major research universities or the wherewithal to build one—are they doomed to be backwaters? Not necessarily. Almost every city has at least one college or teaching hospital, all of which have faculty who could be prospective entrepreneurs with the right support. Investing in their departments is one of the best ways to incubate a startup economy and, potentially, generate economic development. An example is the University of Tulsa which has unique expertise in its computer science department of training students in tech security who fill a pipeline of jobs to the NSA. The university is working to broaden its program so it can also produce students for computer science positions at other public and private employers. Another opportunity for cities is to encourage local philanthropists to found targeted research institutes, as Allen did in Seattle. The founders do not have to be only people who made their wealth in tech businesses.

Some cities are working to make use of technologies for the benefit of the residents while others appear politically opposed or paralyzed. On as simple a matter as implementing a plan for convenient and safe pickup and drop-off zones for rideshares, Seattle sits on proposed plans. At the same time, cities such as Toronto and Columbus, working with businesses and citizen groups, are formulating plans for the integration of autonomous vehicles into their cities' transportation plans. Toronto has adopted an Automated Vehicles Tactical Plan so that Toronto is AV ready in 2022.[10] Columbus, the winner of the U.S. Department of Transportation's Smart City Challenge in 2016, was awarded $50 million provided by Paul Allen's Vulcan. They are using part of the money for the promotion of autonomous vehicles to make Columbus the "most prosperous region in the United States" for people's "economic and social well-being."[11]

Innovations in artificial intelligence, medicine, communications, and transportation technology will revolutionize our lives. In the future, drones will be able to replace many delivery trucks, and self-driving cars will no longer be a cartoon fantasy from the Jetsons. Biotech will cure a wide range of diseases.

In the next ten years, I am convinced that we will ride to work in driverless vehicles of all sorts. At lunchtime, we might browse a store and, rather than waiting in line at the check-out counter, we'll pay with our phones. If an item is out of stock, it will be delivered by drone, arriving by the time we get home. After work, we might take a self-driving taxi to dinner at a restaurant where automatons in the kitchen have chopped the vegetables— delivered by autonomous vehicles. When we're ready for the check, instead of waving down a waiter and scrutinizing a small piece of paper in semi-darkness, we'll pay by phone.

Technology can change crime detection with facial-recognition software and will upend healthcare through targeted therapies and programs that can analyze our symptoms and medical histories so accurately we often won't have to arrange for a doctor's appointment.

One way or another, most of the technologies transforming our lives use artificial intelligence. Prosaically, we see AI at work when Amazon analyzes your purchasing patterns to predict what products you might like to buy next. Or when Pandora creates musical selections based on your "thumbs up" or "thumbs down" on the songs it plays. The technology is not yet foolproof. But on many complex tasks, whether in reducing auto accidents, identifying people in photos, or finding cancer in MRI scans, AI promises to do a better job than humans.

I believe transportation will be the first area where the general public will notice their lives being changed by AI—specifically, in autonomous vehicles. Taking advantage of artificial intelligence smart vision for autonomous vehicles offers the promise of cleaner air (plus the benefit of more AVs being electric), fewer auto accidents (with the elimination of human error), and more room for public parks. As described earlier, AI can analyze video images to distinguish between a roadside sign and a person on a bicycle—essential for autonomous driving.

Imagine hopping into a self-driving Uber or Lyft-like service that you've summoned through your smartphone to take you to a transit terminal. You will no longer need to drive into town and pay for parking in a dim garage—an obvious benefit to anyone who lives far from public transit but works in a city, as well as those who have difficulty driving.

Street-parking spaces could then be reclaimed for bikes and pedestrians, encouraging healthier lifestyles.

The replacement of gas-fueled cars with electric vehicles is indeed happening more slowly than many expected, mostly due to their higher cost and limited battery life. All major auto companies have announced ambitious plans for rolling out and marketing numerous electric vehicles, highlighted by GM's announcement that all of its vehicles will be electric by 2035, and there are numerous new tech entrants, including Amazon, that are well-capitalized. Electric technologies continue to improve, and increased production will lower prices to less than combustion engine vehicles. Compared with the cost of purchasing a car maintaining, operating, and insuring it—using autonomous vehicle services could save us up to 50 percent annually.

In the not-too-distant future, I expect to see portions of our interstate highways reserved exclusively for autonomous buses and trucks during certain periods. Drones will likely handle many deliveries, flying them through the air to land near our homes or on office rooftops. Drones can fly into buildings through an open window, creating mini-airports with drones delivering products more quickly and cheaply than human-operated freight. The necessary technology in GPS, smart vision, and object avoidance is ready. For the past five years, Amazon has been developing its drones in a building in South Seattle, reconfigured to create an indoor flight-test space. Meanwhile, Google is already piloting a drone project in Australia to test the machines' reliability for delivering take-out food from restaurant kitchens to suburban homes.

The most successful cities of the future will be those that are best able to use their enhanced creativity and wealth to solve their social problems while continuing to foster that creativity and wealth. This can best be accomplished if local governments and companies work together smartly using new technologies. But if cities treat businesses as the enemy and businesses fail to engage their money and talents, I fear the centers of our cities may revert to the conditions of the 1960s and 1970s with empty buildings, economically depressed, and worsening social problems with most productive and creative residents having fled to the suburbs and beyond.

Governments can slow the implementation of new technologies, but in my view, the benefits of these technologies in improving everyone's lives and solving civic problems are so powerful that in the long term they likely will overwhelm attempts to stifle them. If activists can block or delay new technologies, such as the use of artificial intelligence in law enforcement and autonomous vehicles, it may be that the successful regions of the future will be suburban areas and cities and towns willing to work with businesses to apply these technologies to civic problems.

Tech companies are already showing that with their technologies, management skills, and scale, they can provide significant help in solving local and global problems. Individuals like Gates and Bezos also have the resources, skills, and commitments to help solve these problems. In its twenty-year history, the $50 billion Gates Foundation has made enormous contributions to global health and other issues. Expect its next twenty years to be equally impactful.

Likewise, expect that Bezos's work during the next twenty years will be devoted to helping solve some of our most pressing problems. In the last couple of years, he has already established a $10 billion Earth Fund for helping solve global warming and provided $1 billion each to homelessness and early childhood education. Gates is known as much for his most recent twenty years leading his foundation as his first twenty years creating and growing Microsoft. Expect the same with Bezos. Twenty years from now, he will be as well known for his second twenty years as his first.

The best outcome for cities, large and small, is for them to work together with businesses and individuals to energize their economic flywheels for providing innovation, jobs, and wealth, and to strengthen their livability flywheels for solving problems of education, public safety, segregation, homelessness, housing, and transportation. Businesses, nonprofits, and individuals can do a lot, but ultimately successful cities will also depend on the political decisions of city governments and their voters.

Nothing should be more exciting than the idea of aggressively applying our human ingenuity to improve our world.

ACKNOWLEDGMENTS

Where to begin in acknowledging people who contributed to this book? It is a lifetime of family, friends, business partners, political associates, and thinkers.

Many people have had older mentors who helped them, and indeed I too have worked for people who set high standards for me and provided a moral compass. But most of my inspiration has come from younger people who provided new ideas, wit, and energy.

My wife Judi Beck has supported me in every way possible as I made my laptop part of our family. My children (Robert, Katherine, John, Carson, and Jessica) and now grandchildren (Charlie, Vivian, Ian, and Abby) have inspired me to want to help make this a better world. My daughter Jessica, a poet, vigorously edited one of my chapters.

There is no doubt that this book would not have happened without the interest and work of Greg Shaw—friend, publisher, agent, editor, and advisor. He was ably assisted by Claudia Rowe who provided me erudite and essential editing. Jim Levine, my agent at Levine Greenberg Rostan, gave me valuable early encouragement and advice and worked tirelessly to interest publishers. And thanks to Columbia University Press, who is willing to publish a book that is out-of-step with much of present-day currents.

Harvard College and Columbia Law School introduced me to new ideas and intellectual rigor, even though I benefited as much from my fellow students as the faculty. In my earlier career as a lawyer at Cravath, Swaine & Moore and Perkins Coie, I benefited greatly from several of my senior partners who did not allow sloppy thinking or writing (Bill Marshall, Harry Riordan, John Ellis, Harold Olsen, Paul Coie, and Doug Beighle) and from younger lawyers who kept me on my toes (Stewart Landefeld, Michelle Wilson, Margaret McKeown, Chuck Katz, and many others). And, of course, Bill Gates Senior, who became a life-long friend.

I have been fortunate to have worked for, learned from, and been inspired by two of the most innovative business leaders of our time: Craig McCaw and Jeff Bezos. I have learned so much from everyone at Madrona Venture Group who took our original idea of investing in early-stage tech companies and built a firm far more impactful than I imagined. I continue to learn from them every day. Paul Goodrich, Matt McIlwain, Tim Porter, Scott Jacobson, Len Jordan, Soma Somasegar, Steve Singh, Hope Cochran, and Troy Cichos. Bill Ruckelshaus and Jerry Grinstein were compelling role models.

I have drawn knowledge and inspiration from dozens of technologists over many years, such as futurist and prolific writer, thinker, and author George Gilder, who first became a friend in college; and polymaths Craig Mundie, Oren Etzioni, Ed Lazowska, and Carver Mead. John Stanton, the builder of the third-most-important wireless service company, T-Mobile, who as a younger colleague taught me the ropes in my first days at McCaw Cellular, always saw the big picture and became a friend and partner in business, political, and nonprofit endeavors.

I have shared political thinking and adventures with a wide spectrum of political thinkers who are also doers. Bruce Chapman, my college room-mate, and I have discussed, debated, and spent endless hours working on political issues and causes. Bruce and I were part of a group of young Turks who helped awaken Seattle politics in the 1970s as he was elected to the Seattle City Council and Washington State secretary of state, and Chris Bayley defeated the longtime incumbent prosecuting attorney. Paul Schell, mayor of Seattle, and long-time friend and fellow activist, sadly deceased. David Brewster, thinker and writer on all things political and the arts, most recently during COVID has been a walking partner where we discuss the political state of the city and the world.

A list too long to cite their many accomplishments and their help: Chris Gregoire, Brad Smith, Mary Snapp, Dan Li, John Doerr, Bill Gurley,

Patty Stonesifer, Steve Duzan, Chris Diorio, Carver Mead, Bob Nelsen, Greg Gottesman, Jeff Wilke, Jeff Blackburn, David Zapolsky, Paula Reynolds, Keith Vernon, Mary Alberg, John Cook, John Oppenheimer, Steve Buri, Cindy Petek, Roberta Katz, Margaret O'Mara, Paul Aslanian, and Jonathan Sposato.

NOTE ON SOURCES

Sources used in the preparation of this book include personal experiences, interviews, conversations, meetings, calls, correspondence (letters, emails and text), documents, and records. Conversations are based on the author's recollections, notes, and correspondence. For research reports, see specific citations in the book. Newspaper and magazine reporting was used to identify particular facts, dates, and interviews of business and political leaders.

Books

Collins, Jim. *Good to Great: Why Some Companies Make the Leap . . . And Others Don't.* New York: Harper Business, 2001.

Cornett, Mick. *The Next American City: The Big Promise of Our Midsize Metros.* New York: Putnam, 2018.

Emanuel, Rahm. *The Nation City: Why Mayors Are Now Running the World.* New York: Knopf, 2020.

Feld, Brad. *Startup Communities: Building an Entrepreneurial Ecosystem in Your City,* 2nd ed. Hoboken, NJ: Wiley, 2020.

Florida, Richard. *The New Urban Crisis: How Our Cities Are Increasing Inequality, Deepening Segregation, and Failing the Middle Class—and What We Can Do About It.* New York: Basic Books, 2017.

——. *The Rise of the Creative Class—Revisited: Revised and Expanded.* New York: Basic Books, 2019.

Galloway, Scott. *Post Corona: From Crisis to Opportunity.* New York: Portfolio, 2020.

Gilder, George. *Life After Google: The Fall of Big Data and the Rise of the Blockchain Economy.* Washington, DC: Regnery, 2018.

——. *Life After Television.* New York: Norton, 1994.

Gilder, Louisa. *The Age of Entanglement: When Quantum Physics Was Reborn.* New York: Random House, 2008.

Graham, Steve. *Invisible Ink: Navigating Racism in Corporate America.* North Charleston, SC: CreateSpace, 2017.

Green, Ben. *The Smart Enough City: Putting Technology in Its Place to Reclaim Our Urban Future.* Cambridge, MA: MIT Press, 2020.

Hwang, Victor, and Greg Horowitt. *The Rainforest: The Secret to Building the Next Silicon Valley.* Los Altos Hills, CA: Ragenwald, 2012.

Noah Harari, Yuval. *Homo Deus: A Brief History of Tomorrow.* New York: Harper, 2017.

Isaacson, Walter. The Code Breaker: Jennifer Doudna, Gene Editing, and the Future of the Human Race. New York: Simon & Schuster, 2021.

Isaacson, Walter, and Jeff Bezos. *Invent and Wander: The Collected Writings of Jeff Bezos, With an Introduction by Walter Isaacson.* Brighton, MA: Harvard Business Review Press, 2020.

Katz, Bruce, and Jeremy Nowak. *The New Localism: How Cities Can Thrive in the Age of Populism.* Washington, DC: Brookings Institution Press, 2018.

Madrona Venture Group. *Madrona Venture Group—25 Years of Innovation (1995–2000).* Seattle, WA: Madrona Venture Group, 2020.

Mazzucato, Mariana. *The Entrepreneurial State: Debunking Public vs. Private Sector Myth.* New York: Public Affairs, 2015.

McClelland, Cary. *Silicon City—San Francisco in the Long Shadow of the Valley.* New York: Norton, 2018.

Mooney, Fred. *Seattle and the Demons of Ambition: From Boom to Bust in the Number One City of the Future.* New York: St. Martin's, 2004.

Moretti, Enrico. *New Geography of Jobs.* Boston: Houghton Mifflin Harcourt, 2012.

Nadella, Satya. *Hit Refresh: The Quest to Rediscover Microsoft's Soul and Imagine a Better Future for Everyone.* New York: Harper Business, 2017.

O'Mara, Margaret. *Cities of Knowledge: Cold War Science and the Search for the Next Silicon Valley (Politics and Society in Modern America).* Princeton, NJ: Princeton University Press, 2004.

——. *The Code: Silicon Valley and the Remaking of America.* New York: Penguin, 2019.

Perkins Coie. *Perkins Coie—A Century of Service.* Seattle, WA: Perkins Coie, 2012.

Rae, Douglas W. *City: Urbanism and Its End.* New Haven, CT: Yale University Press, 2003.

Ridley, Matt. *How Innovation Works: And Why It Flourishes in Freedom.* New York: Harper, 2020.

Saxenian, Anna Lee. *Regional Advantage: Culture and Competition in Silicon Valley and Route 128*. Cambridge, MA: Harvard University Press, 1994.

Smith, Brad. *Tools and Weapons: The Promise and the Peril of the Digital Age*. New York: Penguin, 2019.

Sposato, Jonathan. *Better Together: 8 Ways Working with Women Leads to Extraordinary Products and Profits*. Hoboken, NJ: Wiley, 2017.

Weiss, Mitchell. *We the Possibility: Harnessing Public Entrepreneurship to Solve Our Most Urgent Problems*. Brighton, MA: Harvard Business Review Press, 2021.

NOTES

Prelude to Jeff Bezos's Day 1

1. "World's Most Admired Companies," Fortune.com, accessed May 11, 2021, https://fortune.com/worlds-most-admired-companies/.

2. Mike Baker, "Seattle's Virus Success Shows What Could Have Been," *New York Times*, March 11, 2021, https://www.nytimes.com/2021/03/11/us/coronavirus-seattle-success.html; Erika Fry, "Saving a City: How Seattle's Corporate Giants Banded Together to Flatten the Curve," Fortune.com, April 17, 2020, https://fortune.com/longform/coronavirus-seattle-flatten-curve-amazon-microsoft-starbucks-nordstrom-costco-covid-19-outbreak/?swaswa.

3. On March 6, at 2:43 p.m., the health officer for Public Health—Seattle & King County, . . . sent an email to a half-dozen colleagues, saying, "I want to cancel large group gatherings now. . . . The county's numbers—10 known deaths and nearly 60 confirmed cases as of late morning—were bad and getting worse." David Gutman, Lewis Kamb, and Ken Armstrong, "And the Team Played On," *Seattle Times*, August 24, 2020, https://www.seattletimes.com/seattle-news/times-watchdog/why-the-seattle-sounders-game-went-on-despite-coronavirus-emergency/.

4. Washington State Report Card OSPI, accessed May 11, 2021, https://washingtonstatereportcard.ospi.k12.wa.us/ReportCard/ViewSchoolOrDistrict/100229.

5. Seattle was the center of grunge music with bands Nirvana, Pearl Jam, Sound Garden, and Alice in Chains. Curt Corbain, Nirvana's lead singer, lived with his wife Courtney Love in my Seattle neighborhood, where he committed suicide in 1994.

6. David Cummings, "Employease: A SaaS Pioneer," Entrepreneurship Blog, August 26, 2009, https://davidcummings.org/2009/08/26/employease-a-saas-pioneer/.

7. Jeffrey Bezos, *Amazon Business Plan* (Seattle: Amazon.com, 1995), 24.

8. Bezos, *Amazon Business Plan*, 28.

9. Walter Isaacson in the introduction to his recent book with Bezos compares Bezos to da Vinci, Einstein, and others, because of his innovator nature. "All were very smart. But that's not what made them special. Smart people are a dime a dozen and often don't amount to much. What counts is being creative and imaginative. That's what makes someone a true innovator." Among the common traits Isaacson cites are a passionate curiosity, a "reality-distortion field," and the ability to retain a "childlike sense of wonder." Walter Isaacson, *Introduction to Invent and Wander: The Collected Writings of Jeff Bezos, by Jeff Bezos* (Cambridge, MA: Harvard Business Review Press, 2020), 1–3.

1. Opportunities and Challenges of Cities

1. Jacob Colker and Oren Etzioni, "Analysis: Seattle Startup Ecosystem Poised for Unprecedented Acceleration of Company Creation," *Geekwire*, October 18, 2019, https://www.geekwire.com/2019/analysis-seattle-startup-ecosystem-poised-unprecedented-acceleration-company-creation/.

2. Officially, Amazon selected Crystal City, Virginia, a suburb of Washington, D.C., as its HQ2, with a projection of 25,000 employees. Jonathan O'Connell and Robert McCartney, "It's Official: Amazon Splits Prize Between Crystal City and New York," *Washington Post*, November 13, 2018, https://www.washingtonpost.com/local/amazon-hq2-decision-amazon-splits-prize-between-crystal-city-and-new-york/2018/11/13/d01ec4de-e76e-11e8-b8dc-66cca409c180_story.html.

3. Wired Staff, "May 26, 1995: Gates, Microsoft Jump on 'Internet Tidal Wave,'" *Wired*, May 26, 2010, https://www.wired.com/2010/05/0526bill-gates-internet-memo/.

4. Robert Strohmeyer, "The 7 Worst Tech Predictions of All Time," PCWorld, December 31, 2008, https://www.pcworld.com/article/155984/worst_tech_predictions.html. Some are skeptical that Watson ever made that prediction, but when computers cost millions and filled entire rooms, it should not be surprising that experts doubted there would be dozens, let alone, tens of millions of computers. "Urban legend: I think there is a world market for maybe five computers," GeekHistory, https://geekhistory.com/content/urban-legend-i-think-there-world-market-maybe-five-computers.

5. AnnaLee Saxenian, *Regional Advantage: Culture and Competition in Silicon Valley and Route 128* (Cambridge, MA: Harvard University Press, 1996); Margaret O'Mara, *The Code: Silicon Valley and the Remaking of America* (New York: Penguin, 2019); Margaret Pugh O'Mara, *Cities of Knowledge: Cold War Science and the Search for the Next Silicon Valley* (Princeton, NJ: Princeton University Press, 2015).

6. The principal designer of the Mark I computer, Howard Aiken of Harvard, estimated in 1947 that six digital computers would be sufficient to satisfy the computing needs of the entire United States. John Kopp, "An Illustrated History of Computers, Part 3," 2002, http://www.computersciencelab.com/ComputerHistory/HistoryPt3.htm.

As a Harvard freshman in 1958, I visited Professor Aiken in his office at the urging of his daughter, Kathy Aiken, one of my high school friends in Seattle. Aiken, presumably correctly judging me not likely to be a computer theorist, didn't invite me home for dinner.

7. AT&T's consultants noted all the problems with the early cellphones—the handsets were absurdly heavy, the batteries kept running out, the coverage was patchy, and the cost per minute was exorbitant. As a result, AT&T decided not to pursue the cellular market and gave to the Baby Bells the opportunity to acquire all the cellular markets in the United States for free. This would cost AT&T $18 billion when they had to acquire McCaw Cellular and LIN Broadcasting in 1995 to obtain a national cellular network. "Cutting the Cord," *Economist*, special report, October 7, 1999, https://www.economist.com/special-report/, 999/10/07/cutting-the-cord.

8. Robert J. Gordon, *The Rise and Fall of American Growth: The U.S. Standard of Living since the Civil War* (Princeton, NJ: Princeton University Press, 2017), 12; Benjamin Wallace-Wells, "The Blip," *New York Magazine*, July 19, 2013, https://nymag.com/news/features/economic-growth-2013-7/.

9. "Job market remains tight in 2019, as the unemployment rate falls to its lowest level since 1969," Monthly Labor Review, U.S. Bureau of Labor Statistics, April 2020, https://www.bls.gov/opub/mlr/2020/article/job-market-remains-tight-in-2019-as-the-unemployment-rate-falls-to-its-lowest-level-since-1969.htm.

10. George Gilder, "The Laws of Money," *Gilder Press*, July 28, 2020, https://gilderpress.com/2020/07/28/the-laws-of-money/.

11. Coined by George Gilder in his book *Life After Television: The Coming Transformation of Media and American Life* (New York: Norton, 1990), 23; George Gilder, *Life After Google: The Fall of Big Data and the Rise of the Blockchain Economy*, (Washington, D.C.: Regnery Gateway, 2018), xiv.

12. Enrico Moretti, *The New Geography of Jobs* (Boston: Houghton Mifflin Harcourt, 2013).

13. Moretti, *The New Geography of Jobs*, 23.

14. Richard Florida, *The Rise of the Creative Class* (New York: Basic, 2019).

15. Richard Florida, "Foreword," Mick Cornett and Jayson White, *The Next American City: The Big Promise of Our Midsize Metros* (New York: G.P. Putnam's Sons, 2018), x–xi.

Florida also now declares "The pandemic caused big changes in how we work, and the geography of where we work," Mr. Florida said. "The office as we know it, a space to work, is dead." Keith Schneider, "Local Alliances Put Some Cities on the Fast Track to Recovery," *New York Times*, April 6, 2021, https://www.nytimes.com/2021/04/06/business/midsize-city-pandemic-recovery.html.

The quote "central crisis of our time" comes from Richard Florida, *The New Urban Crisis: How Our Cities Are Increasing Inequality, Deepening Segregation, and Failing the Middle Class—and What We Can Do About It* (New York: Basic Books, 2017), xx.

16. O'Mara, *Cities of Knowledge: Cold War Science and the Search for the Next Silicon Valley*.

17. Staff, "Deal Book," *New York Times*, May 14, 2020, https://www.nytimes.com/2020/05/14/business/dealbook/satya-nadella-microsoft.html.

18. Rachel Lerman and Elizabeth Dwoskin, "Facebook will now let some employees work from anywhere, but their paychecks could get cut," *Washington Post*, May 21, 2020, https://www.washingtonpost.com/technology/2020/05/21/facebook -permanent-remote-work/.

19. Jonathan Miller, "Have You Heard of a 'Co-primary Home?,'" podcast, https:// www.bloomberg.com/news/audio/2020-06-15/have-you-heard-of-a-co-primary-home -podcast.

20. Salesforce, News and Insights, "Our Global Return to the Office: The Salesforce Approach," April 12, 2021, https://www.salesforce.com/news/stories/global -return-to-the-office-the-salesforce-approach/.

21. Amazon News, March 30, 2021, https://www.aboutamazon.com/news/company -news/amazons-covid-19-blog-updates-on-how-were-responding-to-the-crisis#covid -latest.

22. Tulsa Remote, accessed April 27, 2021, https://tulsaremote.com/.

23. LinkedIn Workforce Report | United States | March 2021, March 4, 2021, https:// economicgraph.linkedin.com/resources/linkedin-workforce-report-march-2021.

24. Moretti, *New Geography of Jobs*, page 13.

2. Foundations of the Economic Flywheel

1. Annie Dillard, *The Writing Life* (New York: Harper Perennial 2013), 11.

2. In the spring of 2021, Amazon's share price was over $3,000.

3. Jacqueline Doherty, "Amazon.bomb," *Barron's*, May 31, 1999, https://www .barrons.com/articles/SB927932262753284707; and David Streitfeld, "Analysts, Vendors Increasingly Wary About Amazon," *Washington Post*, February 21, 2001, https://www .washingtonpost.com/archive/business/2001/02/21/analysts-vendors-increasingly -wary-about-amazon/82aee135-065e-4ce9-a293-edbbc83a6552/. Streitfeld, now working for the *New York Times*, continues to be a controversial Amazon skeptic and critic.

4. Sharon Boswell and Lorraine McConaghy, "Lights out Seattle," *Seattle Times*, November 3, 1996, https://special.seattletimes.com/o/special/centennial/november /lights_out.html.

5. John Rhea, "The Plight of the Metal Benders," *Military and Aerospace Electronics*, December 1, 1999, https://www.militaryaerospace.com/home/article/16705917/the-plight -of-the-metal-benders.

3. Seattle's Flywheels Begin Spinning

1. "Mr. Adler recounts the best of his deals lovingly, like a jeweler polishing stones. During a recent conversation, the financier, trim despite his fondness for the delicacies of the Four Seasons and the Quilted Giraffe, ticked off some of his favorites. . . . 'I put $178,000 into Advanced Technology Laboratories in December 1977 and sold it to

Squibb in February 1980 for $60 million,' he recalled. 'I made about $2.5 million there.'"
Ann Crittenden, "Venture Capitalist: A Rise to Riches," *New York Times*, January 6,
1981, https://www.nytimes.com/1981/01/06/business/venture-capitalist-a-rise-to-riches
.html. One of his coinvestors in ATL was Henry Burkardt who, with Adler, was one of
the founders of Data General.

2. See the detailed history of Immunex in "History of Immunex Corporation,"
Funding Universe, accessed March 17, 2021, http://www.fundinguniverse.com/company
-histories/immunex-corporation-history/.

3. For a description of their colorful history, see Andrew Pollack, "A Biotech King,
Dethroned," *New York Times*, September 5, 2013, https://www.nytimes.com/2013/09/06
/business/david-blech-a-biotech-king-dethroned.html.

4. I was on the board of Visio from 1993 until we sold it to Microsoft in January
2000 for $1.3 billion just before the market crashed.

5. Lisa Stiffler, "How Paul Brainerd's Extraordinary Career Went from
Revolutionizing Publishing to Empowering Enviros," GeekWire, July 29, 2018, https://
www.geekwire.com/2018/paul-brainerds-extraordinary-career-went-revolutionizing
-publishing-empowering-enviros/.

6. Camp Glenorchy Eco Retreat, https://www.campglenorchy.co.nz/.

7. Christine Y. Chen, "The Man Who Would Save Satellites," *Fortune*, July 8, 2002,
https://archive.fortune.com/magazines/fortune/fortune_archive/2002/07/08/325859
/index.htm.

8. Anthony Ramirez, "A Telephone Visionary Who Is Cutting the Cords for
Consumers," *New York Times*, November 15, 1992, https://www.nytimes.com/1992/11/15
/business/a-telephone-visionary-who-is-cutting-the-cords-for-consumers.html.

9. The year before I joined McCaw it was Washington's third-most-valuable
company at $6.3 billion, just behind Microsoft also at $6.3 billion. Boeing at $16 billion
was number one.

10. Geraldine Fabrikant, "Craig McCaw's High-Risk Phone Bet," *New York Times*,
May 6, 1990, https://www.nytimes.com/1990/05/06/business/craig-mccaw-s-high-risk
-phone-bet.html.

11. Tim Healy, "Working Behind the Scenes—McCaw's Chief Is Leader with Quiet
Strength," *Seattle Times*, August 26, 1990, https://archive.seattletimes.com/archive/?date
=19900826&slug=1089675.

12. Reuters, "Elon Musk's Neuralink shows monkey with brain-chip playing
videogame by thinking," *Business Insider*, April 9, 2021, https://www.businessinsider
.com/elon-musks-neuralink-shows-monkey-with-brain-chip-playing-videogame-by
-thinking-2021-4.

13. AT&T's other diversification efforts at the time, such as the purchase of
computer company NCR, were notably unsuccessful.

14. "The McCaw Mafia," *Forbes*, February 19, 2001, https://www.forbes.com
/forbes/2001/0219/090.html?sh=28e990346484.

15. Renaissance Capital, "TMT SPAC Colicity Prices Upsized $300 Million
IPO," NASDAQ, February 24, 2021, https://www.nasdaq.com/articles/tmt-spac-colicity
-prices-upsized-%24300-million-ipo-2021-02-24.

4. Microsoft and Amazon Innovate to Success

1. "Allen recalled seeing a "gangly, freckle-faced eighth grader edging his way into the crowd around the Teletype, all arms and legs and nervous energy. He had a scruffy-preppy look: pull over sweater, tan slacks, enormous saddle shoes. His blond hair went all over the place." Nick Eaton, "Paul Allen on 13-Year-Old Bill Gates: 'A Budding Entrepreneur,'" Seattle pi (Microsoft Blog), March 30, 2011, https://blog.seattlepi.com/microsoft/2011/03/30/paul-allen-on-13-year-old-bill-gates-a-budding-entrepreneur/.

2. "Twelve-year-old Bill Gates was two years younger than Allen, but the two became fast friends. A year later, Gates was already suggesting they would run their own company one day. In the same year, Intel released one of its most important microprocessors, the Intel 8080. This tiny piece of hardware was so important to Microsoft's early success that the company's first phone number ended in 8080." Josh Davis, "Retrace the Remarkable Life of Paul Allen Through the Technology That Made It Possible," *University of Washington Magazine*, November 2018, https://magazine.washington.edu/feature/paul-allen-living-computers/.

3. Chris Bayley was a close friend from our Harvard days who was challenging the long-time incumbent.

4. Wired Staff, "May 26, 1995: Gates, Microsoft Jump on 'Internet Tidal Wave,'" *Wired*, May 26, 2010, https://www.wired.com/2010/05/0526bill-gates-internet-memo/.

5. "Amazon is guided by four principles: customer obsession rather than competitor focus, passion for invention, commitment to operational excellence, and long-term thinking." E.g., "The Walt Disney Company Uses AWS to Support the Global Expansion of Disney+," Amazon News Releases, April 29, 2021, https://press.aboutamazon.com/news-releases/news-release-details/walt-disney-company-uses-aws-support-global-expansion-disney.

6. Statement by Jeff Bezos to the U.S. House Committee on the Judiciary, July 28, 2020, Amazon.com, https://www.aboutamazon.com/news/policy-news-views/statement-by-jeff-bezos-to-the-u-s-house-committee-on-the-judiciary.

7. "Leadership Principles," Amazon.com, accessed March 21, 2021, https://www.amazon.jobs/en/principles.

8. Jason Aten, "How Amazon's Departing CEO Jeff Bezos Prioritizes His Time According to the 'One-Way-Door' Rule," *Inc.*, February 3, 2021, https://www.inc.com/jason-aten/how-amazons-departing-ceo-jeff-bezos-prioritizes-his-time-according-to-one-way-door-rule.html.

9. Hayley Tsukayama, "What Amazon's Learned from a Decade of Prime," *Washington Post*, February 3, 2015, https://www.washingtonpost.com/news/the-switch/wp/2015/02/03/what-amazons-learned-from-a-decade-of-prime/.

10. "Other observers believe that Amazon's cloud computing platform is a doomed effort, and they view it as similar to the millions of dollars Amazon had invested during the dot-com boom on unprofitable distribution centers." "Amazon Enters the Cloud Computing Business," Stanford University School of Engineering, 2008, https://web.stanford.edu/class/ee204/Publications/Amazon-EE353-2008-1.pdf.

"The Kindle was greeted with skepticism in some quarters when it was first released last year. The idea of e-books had been around for years and never taken hold. The device was expensive ($399, now $359) and clunky compared with smart phones and iPods." Jeffrey A. Trachtenberg and Christopher Lawton, "Amazon's E-Book Gadgets Sell Out After Oprah's Plug; Calculating the Whim Factor," *Wall Street Journal*, December 4, 2008, https://www.wsj.com/articles/SB122834809064877527; Seth Godin, "The Predictable Lifecycle of the Skeptic (or Even Better, Cynic)," Seth's Blog, August 20, 2008, https://seths.blog/2008/08/the-predictable/; Daniel Eran Dilger, "In-depth Review: Can Amazon's Kindle Light a Fire under eBooks?," *Apple Insider*, December 10, 2007, https://appleinsider.com/articles/07/12/10/in_depth _review_can_amazons_kindle_light_a_fire_under_ebooks.

11. In 2007, after three years of research and development by Lab126, Amazon released the Kindle e-reader; in 2011, the Kindle Fire tablet; in 2014, the Fire TV digital media player and the smaller Fire TV Stick; in 2014, the Fire Phone that flopped; and in 2015, the Echo voice command device.

12. Katherine Khashimova Long, "Amazon Launches App-Based Health Care Service for Seattle-Based Employees," *Seattle Times*, February 19, 2020, https:// www.seattletimes.com/business/amazon/amazon-launches-app-based-health-care -service-for-seattle-based-employees/.

13. Amazon News Release, "Amazon.com Announces First Quarter Results," April 29, 2021, https://ir.aboutamazon.com/news-release/news-release-details/2021 /Amazon.com-Announces-First-Quarter-Results/default.aspx.

14. Eugene Kim and Christina Farr, "Inside Amazon's Grand Challenge: A Secretive Lab Working on Cancer Research and Other Ventures," *CNBC*, June 5, 2018, https:// www.cnbc.com/2018/06/05/amazon-grand-challenge-moonshot-lab-google-glass -creator-babak-parviz.html; "Fred Hutch Microbiome Researchers Use AWS to Perform Seven Years of Compute Time in Seven Days," AWS Solutions, 2019, https://aws.amazon .com/solutions/case-studies/fredhutch-case-study/.

5. On the Precipice of the Future

1. Isaac Asimov's three laws of robotics introduced in his 1942 short story "Runaround," in *I, Robot* (New York: Doubleday, 1950), 40.

1. A robot may not injure a human being or, through inaction, allow a human being to come to harm.
2. A robot must obey the orders given it by human beings except where such orders would conflict with the First Law.
3. A robot must protect its own existence as long as such protection does not conflict with the First or Second Laws.

2. Andreas Kaplan, "Artificial Intelligence: Emphasis on Ethics and Education," *International Journal of Swarm Intelligence and Evolutionary Computation* 9, no. 3 (June 17, 2020), https://www.longdom.org/open-access/artificial-intelligence-emphasis -on-ethics-and-education-54298.html.

3. Jeff Bezos, Shareholder Letter 2017, quoted in Brian Roemmele, "What Is Jeff Bezos's 'Day 1' Philosophy?" *Forbes*, April 21, 2017, https://www.forbes.com/sites/quora/2017/04/21/what-is-jeff-bezos-day-1-philosophy/?sh=3448f7ac1052.

These big trends are not that hard to spot (they get talked and written about a lot), but they can be strangely hard for large organizations to embrace. We're in the middle of an obvious one right now: machine learning and artificial intelligence.... At Amazon, we've been engaged in the practical application of machine learning for many years now. Some of this work is highly visible: our autonomous Prime Air delivery drones; the Amazon Go convenience store that uses machine vision to eliminate checkout lines; and Alexa, our cloud-based AI assistant.... But much of what we do with machine learning happens beneath the surface. Machine learning drives our algorithms for demand forecasting, product search ranking, product and deals recommendations, merchandising placements, fraud detection, translations, and much more. Though less visible, much of the impact of machine learning will be of this type—quietly but meaningfully improving core operations.

Jeff Bezos, "2017 Letter to Shareholders," About Amazon, April 18, 2018, https://www.aboutamazon.com/news/company-news/2017-letter-to-shareholders.

4. He told his audience that big data will enable each person to be "completely understood" by artificial general intelligence machines that can produce a computer facsimile of each detail of a single individual. It would be far too complex for human physicians to make sense of. Craig Mundie "Mankind and Machines—Learning to Understand Human Biology, Together" (keynote address, COSM Conference 2019, Bellevue, WA, September 30–October 2, 2019).

5. Yann LeCun, Facebook's director of AI, quoted in Steven Rosenbush, "Facebook AI Chief Pushes the Technology's Limits," *Wall Street Journal*, August 13, 2020, https://www.wsj.com/articles/facebook-ai-chief-pushes-the-technologys-limits-11597334361.

6. During an interview with Nadella, Altman declared, "I think that A.G.I. will be the most important technological development in human history." Cade Metz, "With $1 Billion From Microsoft, an A.I. Lab Wants to Mimic the Brain," *New York Times*, July 22, 2019, https://www.nytimes.com/2019/07/22/technology/open-ai-microsoft.html.

7. Khari Johnson, "OpenAI Debuts Gigantic GPT-3 Language Model with 175 Billion Parameters," VentureBeat, May 29, 2020, https://venturebeat.com/2020/05/29/openai-debuts-gigantic-gpt-3-language-model-with-175-billion-parameters/.

8. From the Wikipedia article on AGI:

- consciousness: To have subjective experience and thought.
- self-awareness: To be aware of oneself as a separate individual, especially to be aware of one's own thoughts.
- sentience: The ability to "feel" perceptions or emotions subjectively.
- sapience: The capacity for wisdom.

Wikipedia, s.v. "artificial general intelligence," last modified March 9, 2021, https://en.wikipedia.org/wiki/Artificial_general_intelligence.

9. Michael Sainato, "Stephen Hawking, Elon Musk, and Bill Gates Warn About Artificial Intelligence," *Observer*, August 19, 2015, https://observer.com/2015/08/stephen -hawking-elon-musk-and-bill-gates-warn-about-artificial-intelligence/; Peter Holley, "Bill Gates on Dangers of Artificial Intelligence: 'I Don't Understand Why Some People Are Not Concerned,' " *Washington Post*, January 29, 2015, https://www.washingtonpost .com/news/the-switch/wp/2015/01/28/bill-gates-on-dangers-of-artificial-intelligence -dont-understand-why-some-people-are-not-concerned/; Catherine Clifford, "Bill Gates: I do not agree with Elon Musk about A.I. 'We shouldn't panic about it,' " *CNBC*, September 25, 2017, https://www.cnbc.com/2017/09/25/bill-gates-disagrees-with-elon -musk-we-shouldnt-panic-about-a-i.html.

10. Christianna Reedy, "Kurzweil Claims That the Singularity Will Happen by 2045," *Futurism*, October 5, 2017, https://futurism.com/kurzweil-claims-that-the -singularity-will-happen-by-2045.

11. George Gilder, *Life After Google* (New York: Gateway, 2018), 98–101.

12. Craig Mundie, personal conversation, April 15, 2021.

13. Walter Isaacson, *The Code Breaker: Jennifer Doudna, Gene Editing, and the Future of the Human Race* (New York: Simon & Schuster, 2021).

14. KING Staff, "Seattle Drivers Wasted Nearly 6 Days Sitting in Traffic Last Year," *KING TV*, February 19, 2019, https://www.king5.com/article/news/local/seattle-drivers -wasted-nearly-6-days-sitting-in-traffic-last-year/281-87b109d5-0960-4c08-b88b -c5cdee746072.

15. Craig Mundie, National Institutes of Health (NIH) Workshop "Harnessing Artificial Intelligence and Machine Learning to Advance Biomedical Research," July 23, 2018, https://datascience.nih.gov/sites/default/files/AI_workshop_report _summary_01-16-19_508.pdf.

16. Leroy Hood, the founder of the Institute for Systems Biology in Seattle, has long been a leading advocate of personalized medicine. He is the inventor of the first high-speed protein sequencer and protein synthesizer, which allowed scientists to characterize a series of new proteins whose genes could then be cloned and analyzed. These were followed by his development of the high-speed DNA synthesizer. Leroy E. Hood, "Lessons Learned as President of the Institute for Systems Biology (2000–2018)," *Science Direct*, February 2008, https://www.sciencedirect.com/science/article/pii /S1672022918300068?via%3Dihub.

17. Taylor Soper, "Well-Funded Stealthy Biotech Startup Nautilus Hires Former Smartsheet, Isilon, GenapSys Execs," GeekWire, January 5, 2021, https://www.geekwire .com/2021/well-funded-stealthy-biotech-startup-nautilus-hires-former-smartsheet -isilon-genapsys-execs/.

18. Institute for Protein Design, University of Washington, https://www.ipd .uw.edu/.

19. Ewen Callaway, " 'It Will Change Everything': DeepMind's AI Makes a Gigantic Leap in Solving Protein Structures, Google's Deep-Learning Program for Determining the 3D Shapes of Proteins Stands to Transform Biology, Say Scientists," *Nature News*, November 30, 2020, https://www.nature.com/articles/d41586-020-03348-4.

20. Emily H. Jung, Alfred Engelberg, and Aaron S. Kesselheim, "Do Large Pharma Companies Provide Drug Development Innovation? Our Analysis Says No," *Stat News*, December 10, 2019, https://www.statnews.com/2019/12/10/large-pharma -companies-provide-little-new-drug-development-innovation/.

21. "Health care . . . is a key focus for the tech industry, as technology is essential to improving patient experiences, achieving better outcomes, promoting wellness, and developing new treatments. These tools are not just found in doctor's offices and hospitals—increasingly, we interact with them through the devices we carry and wear every day." "Tech-Driven Health and Wellness | VR for Patients," TechNet, last updated February 22, 2021, http://technet.org/newsletter /tech-driven-health-and-wellness-vr-for-patients.

22. Tom Simonite and Gregory Barber, "Alphabet's AI Might Be Able to Predict Kidney Disease," *Wired*, July 31, 2019, https://www.wired.com/story/alphabets-ai -predict-kidney-disease.

23. Semonti Stephens, "Apple Announces Three New Apple Watch Health Studies with Big-Name Partners," Apple, updated September 10, 2019, https://www.apple.com /newsroom/2019/09/apple-announces-three-groundbreaking-health-studies/.

24. Geoff Spencer, "AI and Preventative Health Care: Diagnosis in the Blink of an Eye," Microsoft, September 17, 2018, https://news.microsoft.com/apac/features /ai-and-preventative-healthcare-diagnosis-in-the-blink-of-an-eye/.

25. "Their computing power grows exponentially as the number of qubits expands." From Arthur Herman, "The Quantum Computing Threat to American Security," *Wall Street Journal*, November 10, 2019, https://www.wsj.com/articles/the -quantum-computing-threat-to-american-security-11573411715.

26. "AWS Announces New Quantum Computing Service (Amazon Bracket) Along with AWS Center for Quantum Computing and AWS Quantum Solutions Lab," Business Wire, December 2, 2019, https://www.businesswire.com/news/home/20191202005783/en/.

27. Paul Smith Goodson, "Quantum USA vs. Quantum China: The World's Most Important Technology Race," *Forbes*, October 10, 2019, https://www.forbes.com /sites/moorinsights/2019/10/10/quantum-usa-vs-quantum-china-the-worlds-most -important-technology-race/?sh=42052dd572de.

28. "Microsoft, Purdue Collaborate to Advance Quantum Computing," Purdue University, July 3, 2017, https://www.purdue.edu/research/features/stories/microsoft -purdue-collaborate-to-advance-quantum-computing/.

29. Yaacov Benmeleh, Giles Turner, and Matt Day, "Amazon Is Laying the Groundwork for Its Own Quantum Computer," *Bloomberg*, December 2, 2020, https:// www.bloomberg.com/news/articles/2020-12-01/amazon-is-laying-the-groundwork -for-its-own-quantum-computer.

30. Douglas Gantenbein, "Collaboration Between Amazon and UC Berkeley Advances AI and Machine Learning," Amazon Science, July 7, 2020, www.amazon .science/academic-engagements/collaboration-between-amazon-and-uc-berkeley -advances-ai-and-machine-learning.

31. Staff, "Quantum Computers Will Break the Internet, but Only If We Let Them," Rand, April 9, 2020, https://www.rand.org/blog/articles/2020/04/quantum-computers -will-break-the-internet-but-only-if-we-let-them.html.

32. Myerson: "Now the aerospace engineer has a new focus: enlisting construction firms, mining companies, pharmaceutical manufacturers and even the hospitality industry to begin thinking about the role they can play in the economic development of the moon. What we need to live and work in space." From Bryan Bender, "What We're Going to Need to Live and Work in Space," Politico, November 8, 2019, https://www .politico.com/news/2019/11/08/delalune-space-rob-meyerson-q-and-a-066678.

33. Motorola's less ambitious sixty-six-satellite Iridium system filed for bankruptcy in 1998, but today it is one of the few low earth orbiting systems providing satellite service. It does so at a relatively high price and with very low data rates, serving less than eight hundred thousand customers worldwide. Several geosynchronous systems provide very expensive voice and data services to large domed antenna systems. A more recent system that utilizes a flat panel antenna, developed by Kymeta (in which Gates is the primary shareholder), is coming into service, but those antennas, which are designed for yachts and other specialized users, are still quite expensive. Todd Bishop, "Bill Gates Leads $85M Investment in Kymeta as Satellite Broadband Venture Preps for Key Rollout," GeekWire, August 25, 2020, https://www.geekwire.com/2020/bill-gates-leads-85m-investment-kymeta -satellite-broadband-venture-preps-key-rollout/.

34. "UN Broadband Commission Aims to Bring Online the World's 3.8 Billion Not Connected to the Internet," Food and Agriculture Organization of the United Nations, January 24, 2018, http://www.fao.org/e-agriculture/news/un-broadband-commission -aims-bring-online-world%E2%80%99s-38-billion-not-connected-internet.

35. Alan Boyle, "Amazon to Offer Broadband Access from Orbit with 3,236-Satellite 'Project Kuiper' Constellation," GeekWire, April 4, 2019, https://www.geekwire.com /2019/amazon-project-kuiper-broadband-satellite/.

36. Molecular Information Systems Lab (MISL) at the University of Washington, accessed May 1, 2021, https://misl.cs.washington.edu/; Abstract Submission, "Cas9-Mediated Random Access in DNA Data Storage," Nicolas Cardozo, Karen Zhang, Delaney Wilde, Charlie Anderson, Karin Strauss, Luis Ceze, and Jeff Nivala, Synthetic Biology: Engineering, Evolution & Design (SEED), 2021, https://aiche.confex.com/aiche /seed2021/poster/papers/index.cgi?username=633622&password=625312.

37. "Metamaterials: Bending Nature," Intellectual Ventures, January 25, 2017, https://www.intellectualventures.com/buzz/insights/metamaterials-bending-nature.

38. Myhrvold has spun off several companies based on metamaterials. Kymeta has developed a radically new kind of flat satellite antenna for high-speed internet service. Echodyne is building new kinds of scanning radar for site security, drones, and self-driving cars. Other companies include Evolv Technologies and Pivotal Communications. Madrona has invested in Echodyne along with Gates.

6. Investing in the Future: Talent and Capital

1. Steve Jobs was famous for building products that no one had ever purchased before, so there was no way to predict if Apple could sell millions. Similarly, Bezos does not base the launch of a new service or product on a market survey.

2. Madrona Venture Group, "Madrona Venture Group—25 Years of Innovation, 1995–2020," (Seattle: Madrona Venture Group, 2020), 19.

3. National Advisory Council on Innovation and Entrepreneurship, "Report to Secretary Locke: Improving Access to Capital for High-Growth Companies," U.S. Economic Development Administration, June 2011, https://www.eda.gov/files/oie/nacie/NACIE-Report-Access-to-Capital.pdf.

4. Board of Editors, "They Have Very Short Memories," *New York Times*, March 10, 2012, https://www.nytimes.com/2012/03/11/opinion/sunday/washington-has-a-very-short-memory.html; Steven Rattner, "A Sneaky Way to Deregulate," *New York Times*, March 3, 3013, https://opinionator.blogs.nytimes.com/2013/03/03/a-sneaky-way-to-deregulate/.

5. Kurt Badenhausen, "The Top 10 Rising Cities for Startups," *Forbes*, October 1, 2018, https://www.forbes.com/sites/kurtbadenhausen/2018/10/01/the-top-10-rising-cities-for-startups/?sh=362ba9876b37; Dustin McKissen, "How St. Louis Is Redefining Its Economy by Focusing on Startups," *Forbes*, October 1, 2018, https://www.forbes.com/sites/kurtbadenhausen/2018/10/01/the-top-10-rising-cities-for-startups/?sh=62479cd56b37.

6. "PSL Studio," Pioneer Square Labs, last updated February 4, 2021, https://www.psl.com/studio.

7. Alejandro Cremades, "10 Startup Accelerators Based On Successful Exits," *Forbes*, June 7, 2018, https://www.forbes.com/sites/alejandrocremades/2018/08/07/top-10-startup-accelerators-based-on-successful-exits/#709477a34b3b.

8. Amy Nelson, Founder and CEO, "The Riveter," accessed May 1, 2021, https://theriveter.co/voice/author/amy-nelson/.

9. "US Venture Capital Investment Surpasses $130 Billion in 2019 for Second Consecutive Year," Pitchbook, January 14, 2019, https://www.prnewswire.com/news-releases/us-venture-capital-investment-surpasses-130-billion-in-2019-for-second-consecutive-year-300986237.html.

10. Taylor Soper, "Why Rand Fishkin Avoided VC and Made His Investment Documents Public for New Startup SparkToro," GeekWire, June 7, 2018, https://www.geekwire.com/2018/rand-fishkin-avoided-vc-made-investment-documents-public-new-startup-sparktoro/.

11. Doerr's $5 million investment in Netscape Communications reaped $400 million for Kleiner.

12. SEC S-1: 3,573,737 shares outstanding after financing at $14.05 per share equals $50,211,005 post money value.

13 Between December 6, 1995, and May 16, 1996, the company issued 3,021,000 shares to twenty-three investors at $.3333 per share (an aggregate of $1,007,000). After this investment, the postmoney valuation of the company was $5,091,515. "Amazon, Inc., IPO: S-1/A, May 21, 1997," SEC Info, http://www.secinfo.com/dr643.8Yk.htm#54a8.

14. Harvard finally relented and invested in our 2021 fund.

15. John Doerr, email message to Chris Douvos, June 9, 2005.

16. Jeff Bezos, email message to Brian Bank, February 28, 2005.

17. Washington State Investment Board invested $40 million in each of OVP VI and VII.

18. John Cook, "The Mighty Madrona: Why One VC Firm Is So Critical to Seattle (and Why That's a Little Scary)," GeekWire, March 30, 2015, https://www.geekwire .com/2015/the-mighty-madrona-why-one-vc-firm-is-so-critical-to-seattle-and-why -thats-a-little-scary/.

19. Zoë Bernard and Kate Clark, "As Silicon Valley Turns Attention to Race, Black Entrepreneurs Detail Prejudice," Information, June 11, 2020, https://www .theinformation.com/articles/as-silicon-valley-turns-attention-to-race-black -entrepreneurs-detail-prejudice; Nico Grant, "Black Venture Capitalists Confront Silicon Valley's Quiet Racism," *Bloomberg*, August 24, 2020, https://www.bloomberg .com/news/features/2020-08-24/black-venture-capitalists-confront-silicon-valley-s -quiet-racism; "We Rise Together," BLCK VC, accessed March 25, 2021, www.blckvc.com.

20. "eLesson: Unconscious Bias," Microsoft, May 16, 2007, https://www.mslearning .microsoft.com/course/72169/launch.

21. Zillow, "The Board Challenge Launches Pledge for U.S. Boards Of Directors to Add a Black Director Within One Year," *PR Newswire*, September 9, 2020, https:// www.prnewswire.com/news-releases/the-board-challenge-launches-pledge-for-us -boards-of-directors-to-add-a-black-director-within-one-year-301126074.html.

22. Black Boardroom Initiative, accessed May 1, 2021, https://www.blackboard roominitiative.org/.

23. Conversation with Michael Perham, SVP and General Counsel, RealNetworks, April 22, 2021.

24. "Introducing the Talent x Opportunity Fund," Andreessen Horowitz, accessed May 1, 2021, https://a16z.com/2020/06/03/talent-x-opportunity/.

7. New Models for Success: Oklahoma City, Tulsa, and Kansas City

1. Marc Stiles, "Seattleites Don't Realize It Yet, but It's Pronounced Tech-Oma," *Puget Sound Business Journal*, November 17, 2020, https://www.bizjournals.com/seattle /news/2020/11/17/tacoma-is-a-burgeoning-tech-hub.html.

2. A longtime personal friend and advisor, with a lengthy history of accomplishments in the business, non-profit, and literary worlds. https://www.linkedin.com/in /greg-shaw-a6585762/.

3. Jon Marcus, "Small Cities Are a Big Draw for Remote Workers During the Pandemic," NPR, November 16, 2020, https://www.npr.org/2020/11/16/931400786 /small-cities-are-a-big-draw-for-remote-workers-during-the-pandemic?utm_term =nprnews&utm_medium=social&utm_campaign=npr&utm_source=twitter.com&s =03.

4. Mick Cornett, *The Next American City: The Big Promise of Our Midsize Metros* (New York: Putnam's, 2018), 6.

5. Brian Hardzinski, "'Pride of the Plains': National Geographic Calls Oklahoma City 'Best Trip' of 2015," *KGOU*, December 2, 2014, https://www.kgou.org/post/pride -plains-national-geographic-calls-oklahoma-city-best-trip-2015.

6. "About," Francis Tuttle Technology Center, February 10, 2020, https://www
.francistuttle.edu/about.

7. "Kauffman Indicators of Entrepreneurship," 2020, https://indicators.kauffman
.org/.

8. Clifford Krauss, "'I'm Just Living a Nightmare': Oil Industry Braces for
Devastation," *New York Times*, April 21, 2020, https://www.nytimes.com/2020/04/21
/business/energy-environment/coronavirus-oil-prices-collapse.html.

9. "About Us," Oklahoma City National Memorial & Museum, last updated
September 4, 2019, https://memorialmuseum.com/museum/about-us/.

10. Sarah Holder, "Stop Complaining About Your Rent and Move to Tulsa, Suggests
Tulsa," *Bloomberg*, November 16, 2018, https://www.bloomberg.com/news/articles/
2018-11-16/work-remotely-from-tulsa-ok-for-one-year-get-10-000; Don Reisinger, "Tulsa
Will Pay You $10,000 to Move There for a Year," *Fortune*, November 14, 2018, https://
fortune.com/2018/11/14/tulsa-remote-worker-offer/.

11. Zlanti Meyer, "Tulsa wants to pay you $10,000 to move there and work
remotely," Fast Company, October 29, 2019, https://www.fastcompany.com/90423874
/tulsa-wants-to-pay-you-10000-to-move-there-and-work-remotely.

12. Laura Forman, "For Newly Remote Workers, Small Town U.S.A. Will Lose
Its Allure Soon Enough," *Wall Street Journal*, June 19, 2020, https://www.wsj.com
/articles/for-newly-remote-workers-small-town-u-s-a-will-lose-its-allure-soon-enough
-11592559006.

13. Michael Overall, "Tulsa Remote Will Now Help Participants Buy Homes
Here," Tulsa World, February 23, 2021, https://tulsaworld.com/news/local/tulsa-remote
-will-now-help-participants-buy-homes-here/article_37972d76-7530-11eb-a632
-534f39500279.html.

14. Rhett Morgan, "The Holberton School, a Software Engineering Institute,
to Double in Size," Tulsa World, March 21, 2021, https://tulsaworld.com/business
/local/the-holberton-school-a-software-engineering-institute-to-double-in
-size/article_1ae86022-8738-11eb-b475-537be4b0731f.html; Learn to Code in Tulsa,
Holberton, accessed May 1, 2021, https://www.holbertonschool.com/campus_life
/tulsa.

15. Wikipedia, s.v. "Tulsa Race Massacre: Property Losses," last updated February 23,
2021, https://en.wikipedia.org/wiki/Tulsa_race_massacre#Property_losses.

16. "Tulsa Race Riot, a Report by the Oklahoma Commission to Study the Tulsa
Race Riot of 1921," Oklahoma Historical Society, February 28, 2001, https://www
.okhistory.org/research/forms/freport.pdf.

17. "Greenhill Industrial Park," OLT Realty LLC, August 15, 2020, https://www
.greenhilltulsa.com/; Robert Evatt, "Tulsa's New Industrial Park Ready for Business,"
Tulsa World, last updated February 19, 2019, https://tulsaworld.com/business/tulsas
-new-industrial-park-ready-for-business/article_51b6ae38-5d3c-5e0e-b0a2
-ac3d3165c00c.html.

18. Brian Root, "Policing, Poverty, and Racial Inequality in Tulsa, Oklahoma,"
Human Rights Watch, September 11, 2019, https://www.hrw.org/video-photos
/interactive/2019/09/11/policing-poverty-and-racial-inequality-tulsa-oklahoma.

19. Fred Jones, "Tulsa Welcomes New Amazon Fulfilment Center," *Oklahoma Eagle*, August 6, 2020, http://theoklahomaeagle.net/2020/08/06/tulsa-welcomes-new-amazon-fulfilment-center/.

20. Neile Jones, "Tesla Picks Austin but Could Pick Tulsa Down the Road," *KTUL*, July 22, 2020, https://ktul.com/news/local/tesla-picks-austin-today-but-could-pick-tulsa-down-the-road.

21. "About Us," Greenwood Cultural Center, accessed March 26, 2021, https://greenwoodculturalcenter.com/about-us.

22. "World's Greatest Places 2019," *Time*, accessed March 26, 2021, https://time.com/collection/worlds-greatest-places-2019/; Mark Trieglaff, "Gathering Place: A Park for Everyone," National Recreation and Park Association, February 18, 2021, https://www.nrpa.org/parks-recreation-magazine/2021/march/gathering-place-a-park-for-everyone/.

23. Jennifer S. Vey, "Organizing for Success: A Call to Action for the Kansas City Region," *Brookings Institute Report*, 2006, https://www.brookings.edu/wp-content/uploads/2016/06/20060831_kansascity.pdf; Peter J. Eaton, et al., "Prosperity at a Crossroads: Targeting Drivers of Economic Growth for Greater Kansas City," *Brookings Institute Report*, June 13, 2014, https://www.brookings.edu/research/prosperity-at-a-crossroads-targeting-drivers-of-economic-growth-for-greater-kansas-city/.

24. Carol Tice, "Why 10,000 People Have Taken the *Entrepreneur*'s Pledge," *Entrepreneur*, July 26, 2010, https://www.entrepreneur.com/article/218846.

25. "About 1 Million Cups," 1 Million Cups, May 29, 2012, https://www.1millioncups.com/about.

26. "About the Kauffman Indicators," Ewing Marion Kauffman Foundation, last modified November 12, 2020, https://indicators.kauffman.org/about-the-kauffman-indicators.

27. Victor W. Hwang and Greg Horowitt, *The Rainforest: The Secret to Building the Next Silicon Valley* (Los Altos Hills, CA: Regenwald, 2012).

28. Startland News, accessed May 1, 2021, https://www.startlandnews.com/about/.

29. Timothy Williams, "In High-Tech Cities, No More Potholes, but What About Privacy?," *New York Times*, January 1, 2019, https://www.nytimes.com/2019/01/01/us/kansas-city-smart-technology.html.

30. Tony Lystra, "In a Signal That Office Space Still Matters, Amazon Will Expand in Six U.S. Cities," *Puget Sound Business Journal*, August 19, 2020, https://www.bizjournals.com/seattle/news/2020/08/18/amazon-plans-to-grow-office-space-in-six-cities.html.

31. Amazon Staff, "Decentralized Innovation, As Tech Becomes Increasingly Coastal, Amazon Continues to Hire in Tech Hubs Across North America," Amazon, April 29, 2019, https://www.aboutamazon.com/news/job-creation-and-investment/decentralized-innovation.

32. Anthony Spadafora, "Amazon Takes Big Gamble on the Future of the Office," TechRadar, 2020, https://www.techradar.com/uk/news/amazon-takes-big-gamble-on-the-future-of-the-office.

33. Richard Florida and Adam Ozimek, "How Remote Work Is Reshaping America's Urban Geography," *Wall Street Journal*, March 5, 2021, https://www.wsj.com/articles/how-remote-work-is-reshaping-americas-urban-geography-11614960100.

8. Livable Cities

1. Gene Balk, "What Is Middle Class in Seattle? Families Now Earn Median of $121,000," *Seattle Times*, last updated September 15, 2018, https://www.seattletimes.com/seattle-news/data/what-is-middle-class-in-seattle-families-now-earn-median-of-121000/.

2. Staff, "Best Undergraduate Computer Science Programs Rankings," *U.S. News & World Report*, 2021, https://www.usnews.com/best-colleges/rankings/computer-science-overall; Seattle 2010 Budget, City of Seattle, 2010, https://www.seattle.gov/Documents/Departments/FinanceDepartment/10adoptedbudget/INTRODUCTION.pdf; Seattle 2019–20 budget, City of Seattle, 2019, http://www.seattle.gov/Documents/Departments/FinanceDepartment/19adoptedbudget/adoptedbudgetoverview.pdf.

3. Tom Alberg, "Will the Rise of the Eastside Eclipse the Seattle Boom?," Crosscut, February 20, 2019, https://crosscut.com/2019/02/will-rise-eastside-eclipse-seattle-boom.

4. Stephen M. Graham, *Invisible Ink: Navigating Racism in Corporate America* (Scotts Valley: CreateSpace, 2017).

5. Amazon Description of Invisible Ink by Stephen M. Graham, Amazon.com, https://amazon.com/Invisible-Ink-Navigating-corporate-America/dp/1541171179/ref=tmm_pap_swatch_0?_encoding=UTF8&qid=1618349416&sr=8-2.

6. Graham, *Invisible Ink*, 8.

7. Graham, *Invisible Ink*, xiii, 199. Also:

> Without my mentors, primarily Alston J. Shakeshaft in college and Tom A. Alberg when I was a young lawyer, who knows what would have happened to me? It's one of those what-ifs that we can't answer, because it didn't happen. What I can say is that without these two white men taking a sincere and active interest in my future, I doubt that I would have gone to Yale, and I doubt whether I would have made partner at Perkins Coie. (9)

9. Public Safety and Privacy

1. Vianna Davila, "Ballard Rape Case Intensifies Homeless Debate," *Seattle Times*, last updated May 18, 2018, https://www.seattletimes.com/seattle-news/homeless/seattle-homeless-camp-did-not-check-for-warrants-on-a-resident-before-ballard-rape/.

2. Editorial Board, "Seattle's Persistent Crime Problem Demands Change," *Seattle Times*, last updated April 22, 2019, https://www.seattletimes.com/opinion/editorials/seattles-persistent-crime-problem-demands-change/.

3. Chris Ruffo, "The Invisible Asylum," City Journal, https://www.city-journal.org/olympia-washington-mental-hospitals (2021). "In 1962, Washington State had 7,641 state hospital beds [for the mentally ill] for a total population of 2.9 million; today, it has 1,123 state hospital beds for a population of 7.6 million—a 94 percent per-capita reduction."

4. Seven of the nine-member council endorsed a resolution to cut the police budget by 50 percent. Mike Carter and David Gutman, "'There's no plan': Seattle City Council's moves to slash police funding draw criticism," *Seattle Times*, August 17, 2020, https://www.seattletimes.com/seattle-news/theres-no-plan-seattle-city-councils -moves-to-slash-police-funding-draw-criticism/.

5. Steven Greenhouse, "How Police Unions Enable and Conceal Abuses of Power," *New Yorker*, June 18, 2020, https://www.newyorker.com/news/news-desk/how -police-union-power-helped-increase-abuses.

6. Crisis Assistance Helping Out on the Streets (CAHOOTS), August 27, 2020, https://legislativeanalysis.org/crisis-assistance-helping-out-on-the-streets-cahoots/.

7. "What Is CAHOOTS?," White Bird Clinic, September 29, 2020, https:// whitebirdclinic.org/what-is-cahoots/.

8. Trevor Bach, "One City's 30-Year Experiment with Reimagining Public Safety," *U.S. News & World Report*, July 6, 2020, https://www.usnews.com/news/cities /articles/2020-07-06/eugene-oregons-30-year-experiment-with-reimagining-public -safety.

9. "China has the largest monitoring system in the world. There are some 170 million CCTV cameras across the country, and that's tipped to grow more than three-fold with 400 million more set to be installed by 2020." Jon Russell, "China's CCTV Surveillance Network Took Just 7 Minutes to Capture BBC Reporter," TechCrunch, December 14, 2017, https://techcrunch.com/2017/12/13/china-cctv-bbc-reporter/.

10. "It can be used to prevent 'dissidents' or 'undesirables' from travelling by train or plane or to secretly arrest and detain them." Anna Mitchell and Larry Diamond, "China's Surveillance State Should Scare Everyone," *Atlantic*, February 2, 2018, https:// www.theatlantic.com/international/archive/2018/02/china-surveillance/552203/.

11. Russell, "China's CCTV Surveillance Network."

12. Ellen Barry, "From Mountain of CCTV Footage, Pay Dirt: 2 Russians are Named in Spy Poisoning," *New York Times*, September 5, 2018, https://www.nytimes .com/2018/09/05/world/europe/salisbury-novichok-poisoning.html.

13. "Boston's camera saturation is the norm in major cities. Grants from the Justice Department, Department of Homeland Security and other smaller grants to the Port of Boston allowed the city to amass hundreds of cameras since 9/11, both to monitor for national security and to deter street crime." Maggie Clark, "Security Cameras Were Key to Finding Boston Bomber," Pew, April 18, 2013, https:// www.pewtrusts.org/en/research-and-analysis/blogs/stateline/2013/04/18/security -cameras-were-key-to-finding-boston-bombers.

14. "In a study of police cameras in Baltimore, researchers found that after four months of setting up the cameras downtown, crime in that area was down by about 25 percent, according to a study from the Urban Institute." Clark, "Security Cameras Were Key."

15. Annie McDonough, "How New York City Is Watching You: A Look at the City's Surveillance Apparatus and Its 9,000-Plus Cameras," City & State New York, April 29, 2019, https://www.cityandstateny.com/articles/policy/technology/how-new-york-city-is -watching-you.html.

16. "Why Did a Tech Executive Install 1,000 Security Cameras Around San Francisco?," *New York Times*, July 13, 2020, https://www.nytimes.com/2020/07/10/business/camera-surveillance-san-francisco.html.

17. "In a recent partnership with the LAPD, Ring donated Video Doorbells to one out of every 10 homes. In just six months, that neighborhood saw a 55% reduction in burglaries." Heather Kelly, "Amazon's Smart Doorbell Company Wants to Help Stop Neighborhood Crime," *CNN*, May 8, 2018, https://money.cnn.com/2018/05/08/technology/neighbors-ring-app/index.html.

18. Drew Harwell, "Doorbell-Camera Firm Ring Has Partnered with 400 Police Forces Extending Surveillance Concerns," *Washington Post*, August 28, 2019, https://www.washingtonpost.com/technology/2019/08/28/doorbell-camera-firm-ring-has-partnered-with-police-forces-extending-surveillance-reach/.

19. Cade Metz, "Police Drones Are Starting to Think for Themselves," *New York Times*, December 5, 2020, https://www.nytimes.com/2020/12/05/technology/police-drones.html.

20. Rob Walker, "There Is No Tech Backlash," *New York Times*, September 14, 2020, https://www.nytimes.com/2019/09/14/opinion/tech-backlash.html.

21. Amazon Rekognition has helped locate more than one hundred missing children and identify more than five thousand child sex trafficking victims. Kaiser Larsen, "Thorn Collaborates with Amazon Rekognition to Help Fight Child Sexual Abuse and Trafficking," Amazon Blog, August 3, 2018, https://aws.amazon.com/blogs/machine-learning/thorn-partners-with-amazon-rekognition-to-help-fight-child-sexual-abuse-and-trafficking.

22. Thomas J. Prohaska, "Schools Testing Facial Recognition Despite State Warning," *Buffalo News*, June 2, 2019, https://buffalonews.com/2019/06/02/lockport-schools-giving-facial-recognition-system-a-dry-run-despite-state-warning/.

23 Larry Hardesty, "Study Finds Gender and Skin-Type Bias in Commercial Artificial-Intelligence Systems," MIT News, February 11, 2018, https://news.mit.edu/2018/study-finds-gender-skin-type-bias-artificial-intelligence-systems-0212; Matt Wood, "Thoughts on Recent Research Paper and Associated Article on Amazon Rekognition," AWS Machine Learning Blog, January 26, 2019, https://aws.amazon.com/blogs/machine-learning/thoughts-on-recent-research-paper-and-associated-article-on-amazon-rekognition/.

24. Kate Kaye, "Portland Officials Want to Ban Private Use of Facial Recognition Technology, Citing 'Accuracy Problems,'" GeekWire, September 5, 2019, https://www.geekwire.com/2019/portland-officials-want-ban-private-use-facial-recognition-technology-due-accuracy-problems/.

25. Stephanie Condon, Congress: "It's Time for a Time Out" on Facial Recognition, zdnet, May 22, 2019, https://www.zdnet.com/article/congress-its-time-for-a-time-out-on-facial-recognition/; Sophia Ankel and Alba Asenjo, "A School in Sweden Has Been Fined Over $20,000 for Using Facial Recognition Software to Control Student Attendance," *Business Insider*, August 29, 2019, https://www.businessinsider.com/a-school-used-facial-recognition-to-illegally-record-class-attendance-2019-8.

26. Amazon: "Even with this strong track record to date, we understand why people want there to be oversight and guidelines put in place to make sure facial recognition technology cannot be used to discriminate. We support the calls for an appropriate national legislative framework that protects individual civil rights and ensures that governments are transparent in their use of facial recognition technology." Amazon, "Some Thoughts on Facial Recognition Legislation," Amazon News, February 27, 2019, https://www.aboutamazon.com/news/policy-news-views /some-thoughts-on-facial-recognition-legislation.

27. "Amazon's announcement and the letter to Congress raised alarm bells for nonprofit Fight for the Future, which in July launched what it called the first national campaign urging a federal ban on all government use of facial recognition technology." Sean Walton, "Tech and Police Groups Urge Lawmakers Not to Ban Facial Recognition," *Seattle Times*, September 27, 2019, https://www.seattletimes.com/business/technology /tech-and-police-groups-urge-lawmakers-not-to-ban-facial-recognition/.

28. Kashmir Hill, Activists Turn Facial Recognition Tools Against the Police, *New York Times*, October 21, 2021, https://www.nytimes.com/2020/10/21/technology /facial-recognition-police.html.

29. "That sounds like the sort of potentially invasive and controlling technology that would spark a public *outcry*—if not actual marching in the streets, then at least intense pressure on public officials to keep the tech giants in check, and on the tech giants themselves. But as with other unnerving-sounding innovations—self-driving cars, *beacon devices* that let retailers track customers' in-store behavior, delivery drones, cashier-less stores—we mostly just shrug. Or, as in the case of the Ring, actively participate in the aggressive spread of a technology whose potential implications are unclear." Walker, "There Is No Tech Backlash."

30. Facebook, Google Fund Groups Shaping Federal Privacy Debate, *Bloomberg Law*, November 18, 2019, https://news.bloomberglaw.com/privacy-and-data-security /facebook-google-donate-heavily-to-privacy-advocacy-groups; Microsoft Advocates Comprehensive Federal Privacy Legislation, Microsoft News Blog, November 3, 2005, https://news.microsoft.com/2005/11/03/microsoft-advocates-comprehensive-federal -privacy-legislation/.

31. Allysian Finley, "The Making of a DNA Detective: CeCe Moore, an Amateur Genealogist Turned Professional, Helps Police Crack Decades-Old Cases," *Wall Street Journal*, February 15, 2019, https://www.wsj.com/articles/the-making-of-a -dna-detective-11550272449; Neil Vigdor, "Suspect in 1972 Murder Dies in Suicide Hours Before Conviction," *New York Times*, November 9, 2020, https://www.nytimes .com/2020/11/09/us/terrence-miller-suicide.html.

32. Finley, "The Making of a DNA Detective."

33. "Contact Tracing Apps: Which Countries Are Doing What," MedicalXPress, April 28, 2020, https://medicalxpress.com/news/2020-04-contact-apps-countries.html.

34. Donald G. McNeil Jr., "Coronavirus Can Be Stopped, but Only with Harsh Steps, Experts Say," *New York Times*, March 22, 2020, https://www.nytimes.com/2020/03/22 /health/coronavirus-restrictions-us.html.

35. Patrick Howell O'Neill, "Coronavirus Tracing Apps Can Save Lives Even with Low Adoption Rates," *MIT Technology Review*, September 2, 2020, https://www.technologyreview.com/2020/09/02/1007947/coronavirus-contact-tracing-apps-save-lives-low-15-percent-adoption-rates/.

36. "Contact Tracing Is Failing in Many States. Here's Why," *New York Times*, July 31, 2020, https://www.nytimes.com/2020/07/31/health/covid-contact-tracing-tests.html.

37. Sarah Mervosh and Lucy Tompkins, "How Are Americans Catching the Virus? Increasingly, They Have 'No Idea,'" *New York Times*, October 31, 2020, https://www.nytimes.com/2020/10/31/us/coronavirus-transmission-everywhere.html.

38. In April, the Canadian Civil Liberties Association filed a lawsuit against Waterfront Toronto and Canadian government entities, arguing that "the process that resulted in the Quayside agreements was not transparent, reasonable or accountable." Jordan Pearson, "Why Does Google Want to Hand Its Smart City Data to a Third Party 'Civic Data Trust?,'" Vice, October 16, 2018, https://www.vice.com/en_us/article/vbknkj/google-wants-a-civic-data-trust-for-toronto-smart-city-sidewalk-labs.

39. Kaye, "Portland Quietly Launches Mobile Location Data Project."

40. Tiffany Hsu, "Ad Giant Wins Over Disney with Big Data Pitch," *New York Times*, October 15, 2019, https://www.nytimes.com/2019/10/15/business/media/disney-advertising-publicis.html.

41. Kaye, "Portland Quietly Launches Mobile Location Data Project."

42. "Armed with this data, campaigns can track down and segment potential voters based on apps they use and places they have been, including rallies, churches, and gun clubs. In some cases, voters might see an ad on their mobile phone. In others, companies can match data to a specific person, allowing campaigns to determine who gets a fundraising call or a knock on the door." Sam Schechner, Emily Glazer, and Patience Haggin, "Political Campaigns Know Where You've Been. They're Tracking Your Phone," *Wall Street Journal*, October 10, 2019, https://www.wsj.com/articles/political-campaigns-track-cellphones-to-identify-and-target-individual-voters-11570718889.

43. "Seattle is in a crisis of its own making, with soaring crime in parts of the city enabled by lax enforcement and prosecution. . . . Drug addiction is a root cause, but political dysfunction is exacerbating the problem by allowing prolific offenders to repeatedly steal, threaten and attack people with little consequence. This is causing substantial harm, not only to individuals but the city's appeal as a place to raise families, create jobs and provide opportunity." *Seattle Times* Editorial Board, "Seattle's Persistent Crime Problem Demands Change."

10. Homelessness and PreK-12 Education

1. Sydney Brownstone, "As Seattle Officials War Over Future of Homeless Encampments, Frustration Grows Over Lack of Answers," *Seattle Times*, October 3,

2020, https://www.seattletimes.com/seattle-news/homeless/as-seattle-officials-war-over-future-of-homeless-encampments-frustration-grows-over-lack-of-answers/; Scott Greenstone, "Tents in Seattle Increased by More Than 50% After COVID Pandemic Began, Survey Says," *Seattle Times*, April 3, 2021, https://www.seattletimes.com/seattle-news/homeless/tents-in-seattle-increased-by-more-than-50-after-covid-pandemic-began-survey-says/.

2. Deedee Sun, "School Board Members Ask City of Seattle Not to Clear Homeless Camps on School Property," *KIRO TV*, April 7, 2021, https://www.kiro7.com/news/local/school-board-members-ask-city-not-clear-homeless-camps-school property/T2H2WV2PY5FBPGEOIFF5WT5SW4/.

3. Niall McCarthy, "The U.S. Cities with the Most Homeless People," Statista, January 14, 2020, https://www.statista.com/chart/6949/the-us-cities-with-the-most-homeless-people/.

4. Benjamin Maritz and Dilip Wagle, "Why Does Prosperous King County Have a Homelessness Crisis?," McKinsey & Company, January 22, 2020, https://www.mckinsey.com/industries/public-and-social-sector/our-insights/why-does-prosperous-king-county-have-a-homelessness-crisis.

5. Conversation with former city councilperson and interim mayor, Tim Burgess, Seattle, March 23, 2019.

6. Scott Greenstone and Sydney Brownstone, "Seattle City Council Again Attempts to Rein in Team That Clears Homeless Encampments," *Seattle Times*, October 17, 2019, https://www.seattletimes.com/seattle-news/homeless/seattle-city-council-again-attempts-to-rein-in-team-that-clears-homeless-encampments/.

7. "Seattle and King County Create New Unified Regional Homelessness Authority: Evidence-Based, Accountable, and Equitable," King's County (website), December 18, 2019, https://kingcounty.gov/elected/executive/constantine/news/release/2019/December/18-governance-homelessness.aspx.

8. "A 'homelessness authority' was supposed to get Seattle and its suburbs on the same page; after a slow year, they may be further apart. . . . A year ago, King County and city of Seattle signed an agreement to set up a 'regional homelessness authority,' trying to unite the county's major cities to rally against the problem of homelessness. But the authority has yet to really start its work. Scott Greenstone, "A 'Homelessness Authority' Was Supposed to Get Seattle and Its Suburbs on the Same Page; After a Slow Year, They May Be Further Apart," *Seattle Times*, January 4, 2021, www.seattletimes.com%2Fseattle-news%2Fhomeless%2Fa-homelessness-authority-was-supposed-to-get-seattle-and-its-suburbs-on-the-same-page-after-a-slow-year-they-may-be-further-apart.

9. Tim Burgess, "Opinion: End Homeless Tent Encampments with Compassion and Accountability," *Seattle Times*, January 29, 2021, https://www.seattletimes.com/opinion/end-homeless-tent-encampments-with-compassion-and-accountability/.

10. Housing First, National Alliance to End Homelessness, April 20, 2016, https://endhomelessness.org/resource/housing-first/. Chris Rufo makes a strong argument as to why Housing First is not the solution. Chris Rufo, "The Limits of Housing First," April 21, 2021, https://christopherrufo.com/the-limits-of-housing-first/.

11. Mike Lewis, "Initiative seeks to force Seattle to fund homeless housing and then clear camps," GeekWire, April 1, 2021, https://www.geekwire.com/2021/initiative -seeks-force-seattle-fund-homeless-housing-clear-camps/.

12. Gregg Colburn et al., "Impact of Hotels as Non-Congregate Emergency Shelters" (Seattle: University of Washington, 2020), https://regionalhomelesssystem .org/wp-content/uploads/2020/11/Impact-of-Hotels-as-ES-Study_Full-Report_Final -11302020.pdf.

13. Daniel Malone, "Opinion, Reimagining How We House the Homeless Beyond the Shelter Model," Seattle Times, December 11, 2020, https://www.seattletimes.com /opinion/reimagining-how-we-house-the-homeless-beyond-the-shelter-model/.

14. Kim Eckart, "Turning Hotels Into Emergency Shelter as Part of COVID-19 Response Limited Spread of Coronavirus, Improved Health and Stability," University of Washington, October 7, 2020, https://www.washington.edu/news/2020/10/07 /turning-hotels-into-emergency-shelter-as-part-of-covid-19-response-limited-spread -of-coronavirus-improved-health-and-stability/.

15. Ted Land, "King County Plans to Buy a Dozen Hotels to House Homeless in 2021," KING TV, January 1, 2021, https://www.king5.com/article/news/local/homeless /king-co-plans-to-buy-a-dozen-hotels-to-house-homeless-in-2021/281-242108da-7d5b -4021-b14b-38ee71979095; Conor Dougherty, "One Way to Get People Off the Streets: Buy Hotels", New York Times, April 17, 2021, https://www.nytimes.com/2021/04/17 /business/california-homeless-hotels.html.

16. "New McKinsey Report Dives Into Seattle-Area Homelessness Crisis," Seattle City Council Insight, February 4, 2020, https://sccinsight.com/2020/02/04/new -mckinsey-report-dives-into-seattle-area-homelessness-crisis/.

17. Unpublished study, presented to Challenge Seattle Meeting, Boston Consulting Group, September 2019.

18. Taylor Soper and Monica Nickelsburg, "Microsoft Commits $500M to Address Affordable Housing and Homelessness in the Seattle Region," GeekWire, January 16, 2019, https://www.geekwire.com/2019/microsoft-will-spend-500-million-address-affordable -housing-homelessness-seattle-region/.

19. Daniel Beekman and Scott Greenstone, "November Ballot Campaign to Insert Homeless Policy in Seattle Charter Adds Sunset Clause, New Wording on Encampments," Seattle Times, April 17, 2021, https://www.seattletimes.com/seattle -news/politics/campaign-to-put-homeless-policy-in-seattles-charter-adds-sunset -clause-new-language-on-encampments/.

20. Daisuke Wakabayashi and Conor Dougherty, "Google Pledges to Invest $1 Billion to Ease Bay Area Housing Crisis," New York Times, June 18, 2019, https://www .nytimes.com/2019/06/18/technology/google-1-billion-housing-crisis.html; Marc Stiles, "Amazon launches $2B affordable housing fund for Seattle area and 2 other hubs," Puget Sound Business Journal, January 6, 2021, https://www.bizjournals.com/seattle /news/2021/01/06/amazon-announces-2-billion-housing-fund.html.

21. Benjamin Romano, "In Heart of Amazon's Seattle headquarters, a First -of-its-Kind Family Shelter Takes Shape for Mary's Place," Seattle Times, October 11, 2019, https://www.seattletimes.com/business/amazon/in-heart-of-amazons-seattle

-headquarters-a-first-of-its-kind-family-shelter-takes-shape/; Ronald Holden, "Amazon Adding 5 FareStart Restaurants to Its Seattle HQ Campus," *Forbes*, June 11, 2017, https://www.forbes.com/sites/ronaldholden/2017/06/11/amazon-adding-5-fare -start-restaurants-to-its-seattle-hq-campus/?sh=250dee664896.

22. Plymouth Housing, https://plymouthhousing.org/about/history/, accessed May 13, 2021.

23. Adam Brinklow, "UC Berkeley Professor Blames Rent Control for Housing Shortage," Curbed San Francisco, September 5, 2018, https://sf.curbed.com/2018/9/5 /17824038/berkeley-study-rent-control-prop-10-costa-hawkins; and Adam Brinklow, "Why San Francisco Just Voted Against Affordable Housing for Teachers," Curbed San Francisco, July 12, 2019, https://sf.curbed.com/2019/7/12/20691891/san-francisco-teacher -housing-breed-board-supervisors-vote.

24. Enrico Moretti, *New Geography of Jobs* (Boston: Houghton Mifflin Harcourt, 2013); and Enrico Moretti, "Housing Constraints and Urban Misallocation," 2019, https://www.nber.org/system/files/working_papers/w21154/w21154.pdf.

25. See: https://www.housingconnector.com/.

26. Rahm Emanuel, *The Nation City: Why Mayors Are Now Running the World* (New York: Vintage, 2021), 15.

27. Washington State Report Card, Office Superintendent Public Instruction, 2019, https://washingtonstatereportcard.ospi.k12.wa.us/ReportCard/ViewSchoolOrDistrict /100229.

28. William R. Kerr and Frederic Robert-Nicoud, "Tech Clusters" (working paper 20-063, Harvard Business School, Boston, MA, 2019), https://www.hbs.edu/faculty /Publication%20Files/20-063_97e5ef89-c027-4e95-a462-21238104e0c8.pdf.

29. Hannah Furfaro and Katheine Long, "Teaching Changed Almost Instantly Due to Covid 19. How Long Will It Take to Revolutionize Equity in Education?," *Seattle Times*, July 19, 2020, https://www.seattletimes.com/education-lab/teaching-changed -almost-instantly-due-to-Covid-19-how-long-will-it-take-to-revolutionize-equity-in -education/.

30. Mia Tuan, Tammy Campbell, and Denise Juneau, "Invest in a Diverse Teacher Workforce," *Seattle Times*, July 22, 2020, https://www.seattletimes.com/opinion/invest -in-a-diverse-teacher-workforce/.

31. Kate Taylor, "In San Francisco, Virus Is Contained but Schools Are Still Closed," *New York Times*, November 1, 2020, https://www.nytimes.com/2020/11/01/us/san -francisco-coronavirus-schools-reopening.html.

32. Katie Reilly, "As Schools Close Amid Coronavirus Concerns, the Digital Divide Leaves Some Students Behind," *Time*, March 15, 2020, https://time.com/5803355 /school-closures-coronavirus-internet-access/.

33. Alec MacGillis, "The Students Left Behind by Remote Learning," *New Yorker*, September 28, 2020, https://www.newyorker.com/magazine/2020/10/05/the -students-left-behind-by-remote-learning.

34. Natasha Singer, "Learning Apps Have Boomed in the Pandemic. Now Comes the Real Test," *New York Times*, March 22, 2021, https://www.nytimes.com/2021/03/17 /technology/learning-apps-students.html.

35. Peggy Barmore, "COVID-19 Changed Schooling Profoundly—in Some Ways, for the Better," *Seattle Times*, March 21, 2021, https://www.seattletimes.com/education -lab/schooling-has-changed-forever-heres-what-might-stay-after-things-go-back-to -normal/.

36. Claudia Rowe, "Let's Not Squander the Chance to Reimagine Education in Washington State," Crosscut, March 18, 2021, https://crosscut.com/opinion/2021/03 /lets-not-squander-chance-reimagine-education-washington-state.

37. Michael Perrigo, "Google Classroom Plans to Further Transform Distanced Learning with Its 2021 Feature Roadmap," Chrome Unboxed, February 19, 2021, https://chromeunboxed.com/google-classroom-feature-roadmap-2021; Anne Keehn, "Teaching, Learning & Connecting in 2020: Education in an Extraordinary Year," Zoom Blog, December 21, 2020, https://blog.zoom.us/connecting-in-2020-education-in -an-extraordinary-year/.

38. "The World Is Changing. Tulsa's Students Will Be Ready," Tulsa Public Schools, April 16, 2021, https://www.tulsaschools.org/careers/teach/tulsa-beyond.

39. Parent and Family Engagement, *Clayton County Public Schools Website*, April 14, 2021, https://www.clayton.k12.ga.us/departments/federal_programs/parental _involvement.

40. "What Is the TEALS Program?," Microsoft, accessed April 19, 2021, https:// www.microsoft.com/en-us/teals.

41. See: Rainier Scholars Annual Report 2019-2020, Rainier Scholars, https:// www.rainierscholars.org/wp-content/uploads/2020/10/Rainier-Scholars-Annual -Report-2019-2020.pdf.

42. See: Technology Access Foundation Annual Report 2019, Technology Access Foundation, https://techaccess.org/2019-annual-report/.

43. Lola E. Peters, "School Tech Programs Work," Crosscut, February 11, 2019, https:// crosscut.com/2019/02/school-tech-programs-work-seattle-should-fight-get-one-back.

44. See: https://www.seaciti.org/about#what.

45. See: https://code.org/.

46. Hannah Furfaro and Dahlia Bazza, "What's Next for Seattle Schools' Gifted Programs? Here's What We Know So Far," *Seattle Times*, October 22, 2019, https://www .seattletimes.com/education-lab/faq-whats-next-for-seattle-schools-gifted-programs/.

47. Kenneth K. Wong and Francis X. Shen, "Mayoral Governance and Student Achievement How Mayor-Led Districts Are Improving School and Student Per- formance," *American Progress*, March 2013, https://www.americanprogress.org/wp -content/uploads/2013/03/MayoralControl-5.pdf.

48. "50-State Comparison," Education Commission of the United States, accessed April 1, 2021, http://ecs.force.com/mbdata/mbquestNB2C?rep=CS1703.

49. Ashley May, "A Firm Foundation for Charters," *Philanthropy Magazine*, Spring 2018, https://www.philanthropyroundtable.org/philanthropy-magazine/article /a-firm-foundation-for-charters.

50. Robin Lake, Trey Cobb, Roohi Sharma, and Alice Opalka, "Slowdown in Bay Area Charter School Growth: Causes and Solutions," *CRPE*, January 2018, https://www.crpe.org/sites/default/files/crpe-slowdown-bay-area-charter-school

-growth.pdf; Ashley May, "A Firm Foundation for Charters," *Philanthropy Magazine*, Spring 2018, https://www.philanthropyroundtable.org/philanthropy-magazine/article/a-firm-foundation-for-charters.

51. See New York Charter School Center, accessed April 19, 2021, https://nyccharterschools.org/.

52. Dan Taylor, "Best Elementary Schools in DC 2017: Here Are Niche's Top 25 in the District," Patch, February 27, 2017, https://patch.com/district-columbia/washingtondc/best-elementary-schools-dc-2017-here-are-niches-top-25-district; "Waitlist Data," DC Public Charter School Board, April 14, 2020, https://www.dcpcsb.org/evaluating/waitlist-data.

53. "Microsoft Launches Next Stage of Skills Initiative After Helping 30 Million People," Microsoft News Blog, accessed April 19,2021, https://news.microsoft.com/skills/.

54. Tom Clynes, "Peter Thiel Thinks You Should Skip College, and He'll Even Pay You for Your Trouble," *Newsweek*, February 22, 2017, https://www.newsweek.com/2017/03/03/peter-thiel-fellowship-college-higher-education-559261.html.

11. Transportation and Environment

1. INRIX 2020 Global Traffic Scorecard, *Inrix*, accessed April 19, 2021, https://inrix.com/scorecard/.

2. "See How Seattle-Area Commutes Have Changed over the Years—Including the Impact of Covid-19," *Seattle Times*, September 25, 2020, https://projects.seattletimes.com/2020/traffic-lab-data-page/ 2018 statistics most recent available.

3. See: http://inrix.com/.

4. See: https://cascadia.center/aces/.

5. Christopher Elliott, "Are Vehicle Safety Features Actually Reducing Car Accidents?," *Forbes*, September 2, 2020, https://www.forbes.com/advisor/car-insurance/vehicle-safety-features-accidents/; and "Avoiding Crashes with Self-Driving Cars," *Consumer Reports*, February 2014, https://www.consumerreports.org/cro/magazine/2014/04/the-road-to-self-driving-cars/index.htm.

6. Denver: James Gooch, "Uber Reveals Strong Initial Performance of Uber Transit in Denver," Masabi, July 9, 2019, https://www.masabi.com/2019/07/09/uber-reveals-strong-initial-performance-of-uber-transit-in-denver/; and "Uber Collaboration," RTD Masabi, accessed April 1, 2021, https://www.rtd-denver.com/projects/uber-collaboration.

7. Letter from ACES cochairs to Mayor Durkan, May 29, 2020.

8. Michelle Baruchman, "Uber will charge significantly more per trip as new Seattle law goes into effect Jan. 1," *Seattle Times*, December 30, 2020, https://www.seattletimes.com/seattle-news/transportation/embargoed-uber-raising-its-prices-starting-jan-1/.

9. Alexandra Alexa, "Taxis and Ubers in Manhattan Will Get More Expensive as Judge Gives Congestion Fees the Green Light," 6sqft, February 1, 2019, https://www.6sqft.com/taxis-and-ubers-in-manhattan-will-get-more-expensive-as-judge-gives-congestion-fees-the-green-light/.

10. Heidi Groover, "Mayor Durkan proposes 51-cent tax on Uber, Lyft rides in Seattle to fund streetcar, affordable housing," *Seattle Times*, September 18, 2019, https://www.seattletimes.com/seattle-news/transportation/seattle-could-tax-uber-and-lyft-rides-to-fund-downtown-streetcar-affordable-housing/.

11. Aaron Short, "Uber/Lyft Responsible for a Large Share of Traffic," Streetsblog USA, August 7, 2019, https://usa.streetsblog.org/2019/08/07/uberlyft-responsible-for-a-large-share-of-traffic/.

12. Katie Pyzyk, "Uber, Lyft: Ride-hailing Is a Low Contributor to Congestion," Smart Cities Dive, August 6, 2019, https://www.smartcitiesdive.com/news/uber-lyft-ride-hailing-is-a-low-contributor-to-congestion/560261/.

13. Steven Mufson, "General Motors to Eliminate Gasoline and Diesel Light-duty Cars and SUVs by 2035," *Washington Post*, January 28, 2021, https://www.washingtonpost.com/climate-enviro nment/2021/01/28/general-motors-electric/.

14. Nikolaus Lang, Andreas Herrman, Markus Hagenmaier, and Maximilian Richter, "Can Self-Driving Cars Stop the Urban Mobility Meltdown," BCG, July 8, 2020, https://www.bcg.com/publications/2020/how-autonomous-vehicles-can-benefit-urban-mobility.

15. Steven John, "11 Incredible Facts About the $700 Billion US Trucking Industry," Markets Insider, June 3, 2019, https://markets.businessinsider.com/news/stocks/trucking-industry-facts-us-truckers-2019-5-1028248577.

16. NHTSA Releases 2019 Crash Fatality Data, *NHTSA*, December 18, 2020, https://www.nhtsa.gov/press-releases/roadway-fatalities-2019-fars.

17. Bryant Walker Smith, "Human Error as a Cause of Vehicle Crashes," December 18, 2013, Center for Innovation and Society at Stanford Law School, http://cyberlaw.stanford.edu/blog/2013/12/human-error-cause-vehicle-crashes.

18. Tesla Vehicle Safety Report, Accident Data, Q1 2021, Tesla, accessed April 20, 2021, https://www.tesla.com/VehicleSafetyReport.

19. "The Economic and Societal Impact of Motor Vehicle Crashes, 2010 (Revised)," NHTSA, May 2015, https://crashstats.nhtsa.dot.gov/Api/Public/ViewPublication/812013.

20. Tom Alberg and Craig Mundie, "Now Is the Time to Plan for the Autonomous Vehicle Future," Tech Crunch, October 11, 2017, https://techcrunch.com/2017/10/11/now-is-the-time-to-plan-for-the-autonomous-vehicle-future/; Tom Alberg, Craig Mundie, and Daniel Li, Autonomous Vehicle Corridor, September 2017, http://www.madrona.com/wp-content/uploads/2017/09/MVG-I5-Proposal-Digital.pdf; Tom Alberg and Daniel Li, "Opinion, Seattle Should Plan for Driverless Cars, More Ride-sharing," *Seattle Times*, February 9, 2016, https://www.seattletimes.com/opinion/seattle-should-plan-for-driverless-cars-more-ride-sharing/.

21. Moe Vazifeh, "Listen: How Maths Could Drastically Shrink New York's Cab Fleet," interview by Ellie Mackay, *Nature*, May 23, 2018, https://www.nature.com/articles/d41586-018-05249-z.

22. Urban Mobility Report, Texas A&M Transportation Institute, accessed April 20, 2021, https://mobility.tamu.edu/umr/congestion-data/.

23. "Technology and the Future of Cities," Report to the President, February 2016, https://obamawhitehouse.archives.gov/sites/whitehouse.gov/files/images/Blog/PCAST%20Cities%20Report%20_%20FINAL.pdf.

24. A Google-commissioned study predicted that drone deliveries in Canberra could reduce delivery costs for businesses by about $9 million annually. Mike Cherney, "Delivery Drones Cheer Shoppers, Annoy Neighbors, Scare Dogs," *Wall Street Journal*, December 26, 2018, https://www.wsj.com/articles/delivery-drones -cheer-shoppers-annoy-neighbors-scare-dogs-11545843552.

25. Andrew Zaleski, "Cities Seek Deliverance from the E-Commerce Boom," *Bloomberg City Lab*, April 20, 2017, https://www.bloomberg.com/news/articles/2017-04 -20/how-cities-are-coping-with-the-delivery-truck-boom.

26. John Porter, "Amazon's Prime Air Inches Closer to Takeoff in the US with FAA Approval," *The Verge*, August 31, 2020, https://www.theverge.com/2020/8/31/21408646 /amazon-prime-air-drone-delivery-faa-clearance-approval-health-safety-alphabet -wing.

27. Matt Novak, "Alphabet's Wing Lands 'World-First' Approval for Drone Deliveries of Food and Medicine in Australia," Gizmodo, April 9, 2019, https://gizmodo .com/alphabets-wing-lands-world-first-approval-for-drone-del-1833907433.

28. Chase Dowling, Tanner Fiez, Lillian Ratliff, and Baosen Zhang, "How Much Urban Traffic Is Searching for Parking?," *Research Gate*, February 2017, https://www .researchgate.net/publication/313879093_How_Much_Urban_Traffic_is_Searching _for_Parking.

29. Michael Manville and David Shoup, "People, Parking and Cities," *Journal of Urban Planning and Development* 141, no. 4 (December 2005): 233–245, https://www .researchgate.net/publication/235358939_People_Parking_and_Cities.

30. "Solid Advantages," *Woodworks*, accessed April 23, 2021, http://www.fpac.ca /publications/Value-CLT-2011%20NABC%20anaylisfinal.pdf; David Roberts, "The Hottest New Thing in Sustainable Building Is, uh, Wood," Vox, January 15, 2020, https://www .vox.com/energy-and-environment/2020/1/15/21058051/climate-change-building -materials-mass-timber-cross-laminated-clt.

31. Wikipedia, s.v. "environmental impact of concrete," last updated March 17, 2021, https://en.wikipedia.org/wiki/Environmental_impact_of_concrete.

32. Nathaniel Bullard and Kyle Harrison, "Businesses Finally Think Big About Cutting Emissions," Bloomberg, September 27, 2019, https://www.bloomberg.com /opinion/articles/2019-09-27/businesses-finally-think-big-about-cutting-emissions.

33. "Another Green Subsidy Bust," *Wall Street Journal*, December 11, 2020, https:// www.wsj.com/articles/another-green-subsidy-bust-11607730090.

34. Press release, "Amazon Co-founds The Climate Pledge, Setting Goal to Meet the Paris Agreement 10 Years Early," Amazon Press Release Center, September 19, 2019, https://press.aboutamazon.com/news-releases/news-release-details/amazon-co-founds -climate-pledge-setting-goal-meet-paris.

35. In December 2020, Amazon announced that it had become the largest corporate purchaser of renewable energy and by April 2021 had launched 71 utility-scale wind and solar renewable energy projects and 135 solar rooftops on fulfillment centers and sort centers around the globe. Joseph Guzman, "Amazon is scaling up its renewable energy capacity," The Hill, April 19, 2021, https://thehill.com/changing-america/sustainability /energy/549028-amazon-is-scaling-up-its-renewable-energy-capacity.

36. "The Climate Pledge Celebrates passing 100 Signatories," Amazon News Release, April 20, 2021, https://www.aboutamazon.com/news/sustainability/the-climate-pledge-celebrates-surpassing-100-signatories.

37. Sally Jewell, interim CEO of the Nature Conservancy and former director of the Department of the Environment praised Amazon's actions. Ironic for a former outspoken critic of Amazon when she was CEO of REI, for being too aggressive as a competitor and not doing enough to help the communities in which it operated.

38. "Amazon Announces $2 Billion Climate Pledge Fund to Invest in Companies Building Products, Services, and Technologies to Decarbonize the Economy and Protect the Planet," Amazon Press Release Archive, June 23, 2020, https://press.aboutamazon.com/news-releases/news-release-details/amazon-announces-2-billion-climate-pledge-fund-invest-companies.

39. Cities have been outspoken in advocating for action to fight global warming but generally are laggards in purchasing electric buses and government vehicles.

12. Government and Business: Conflicts and Cooperation

1. "The 2006 Annual One Night Count: People Who Are Homeless in King County, Washington," Seattle King County Coalition on Homelessness, January 2006, https://homelessinfo.org/wp-content/uploads/2019/03/2006_ONC_Report.pdf; Homeless Count, King County News Release, July 1, 2020, https://kingcounty.gov/elected/executive/constantine/news/release/2020/July/01-homeless-count.aspx.

2. Louis Aguilar, "Detroit Population Continues to Decline, According to Census Estimate," Bridge Michigan, May 21, 2020, https://www.bridgemi.com/urban-affairs/detroit-population-continues-decline-according-census-estimate; Jon Talton, "Seattle as the Next Detroit? That's Not How We Roll," *Seattle Times*, October 9, 2020, https://www.seattletimes.com/business/seattle-as-the-next-detroit-thats-not-how-we-roll/.

3. Monica Nickelsburg, "Seattle Adopts Minimum Wage for Uber and Lyft Drivers," GeekWire, September 29, 2020, https://www.geekwire.com/2020/seattle-adopts-minimum-wage-uber-lyft-drivers/.

4. Fred Mooney, *Seattle and the Demons of Ambition: From Boom to Bust in the Number One City of the Future* (New York: St. Martin's, 2004), 66.

5. D. J. Wilson, "One Reason the Push Back on the Head Tax Was So Strong," *Washington State Wire*, May 15, 2018, https://washingtonstatewire.com/opposition-to-head-tax/. The October 2019 poll also found that only 19 percent of the voters were more inclined to vote for a candidate who wants to continue the current direction of the council. Crosscut Elway Poll, Seattle City Government 2019, Crosscut, October 2019, https://crosscut.com/sites/default/files/2019-10/crosscutelwaypolloct2019.pdf.

6. The mayor did comment that she often defended Amazon's impact on the city by pointing out its positive economic impact.

7. Danny Westneat, "More Money for Homeless? Seattle Consultants Said No Last Fall," *Seattle Times*, March 10, 2017, https://www.seattletimes.com/seattle-news/politics/more-money-for-homeless-seattle-consultants-said-no-last-fall/; Chris F. Rufo,

"Seattle Under Siege," City Journal, Autumn 2018, https://www.city-journal.org/seattle-homelessness.

8. In a potent sign of the future, Amazon recently decided not to renew an office lease in Seattle that housed one thousand employees.

9. A slowdown in Amazon's Seattle growth may be offset by the growth of the 140 out-of-state companies now located here—unless they are also deterred by Seattle's political climate.

10. Matt Rosenburg, "Cascadia-Arup Report: Deep Bore Tunnels @ $200M-$700M Per Mile," Cascadia Center, November 11, 2008, https://cascadia.center/blog/2008/cascadiaarup_report_tunnels_20/.

11. Editorial Board, "Alaskan Way Viaduct: Recurring Dream," Seattle PI, December 31, 2008, https://www.seattlepi.com/local/opinion/article/Alaskan-Way-Viaduct-Recurring-dream-1296360.php; Mike Wussow, "Cascadia Center, Experts Say Tunnel Costs for Replacing Viaduct a Myth, Tunnel Pros Urge State, County, City to Weigh Real Costs as Decision Near," Cascadia Center, https://www.discovery.org/a/8231/; Mike Wussow, "It's Done: With Pen to Paper, Gregoire Gives Seattle a Tunnel," Cascadia Center, May 12, 2009, https://cascadia.center/blog/2009/it_is_done_with_pen_to_paper_g/.

12. "Business Roundtable Redefines the Purpose of a Corporation to Promote 'An Economy That Serves All Americans,'" Business Roundtable, August 19, 2019, https://www.businessroundtable.org/business-roundtable-redefines-the-purpose-of-a-corporation-to-promote-an-economy-that-serves-all-americans.

13. Editorial Board, "King Warren of the Roundtable," Wall Street Journal, October 6, 2019, https://www.wsj.com/articles/king-warren-of-the-roundtable-11570395953.

14. Big companies, despite the resources they have to devote to local problems, are limited in the time and energy they can focus on public issues. As one CEO pointed out to me, though they need to participate in solving local problems, they can't forget that small innovative companies spend every single day working to satisfy customers and displace the big guys.

15. The Council of Institutional Investors argued that "it is government, not companies, that should shoulder the responsibility of defining and addressing societal objectives." Elizabeth Dilts, "Top U.S. CEOs Say Companies Should Put Social Responsibility Above Profit," Reuters, August 19, 2019, https://www.reuters.com/article/us-jp-morgan-business-roundtable/top-u-s-ceos-say-companies-should-put-social-responsibility-above-profit-idUSKCN1V91EK.

16. Marc Benioff, "We Need a New Capitalism," New York Times, October 14, 2019, https://www.nytimes.com/2019/10/14/opinion/benioff-salesforce-capitalism.html.

17. Michael Liedtke and Barbara Ortutay, "Crusading Tech Mogul Aims to Prove CEOs Can Be Activists Too," ABC News, October 4, 2019, https://abcnews.go.com/Lifestyle/wireStory/crusading-tech-mogul-aims-prove-ceos-activists-66068641.

18. Scott Galloway, Post Corona: From Crisis to Opportunity, 2020 (New York: Penguin Random House), 69.

19. Jeff Bezos, Amazon shareholder letter, accessed April 1, 2021, https://www.sec.gov/Archives/edgar/data/1018724/000119312520108427/d902615dex991.htm.

20. Sandi Doughton, "Fred Hutch Study Targets Front-Line Workers Most At-Risk from the Coronavirus," *Seattle Times*, May 16, 2020, https://www.seattletimes.com /seattle-news/health/fred-hutch-study-targets-front-line-workers-most-at-risk-from -the-coronavirus/.

21. Tom Alberg and Raj Singh, "Coronavirus: Now More Than Ever, We Need to Channel Our Historic 'Seattle Spirit,'" *Seattle Times*, March 23, 2020, https://www .seattletimes.com/opinion/coronavirus-now-more-than-ever-we-need-to-channel-our -historic-seattle-spirit/; Patti Payne, "Coalition of Settle-Area Leaders Raises $27M to Aid Individuals and Businesses," Bizjournals, March 23, 2020, https://www.bizjournals .com/seattle/news/2020/03/23/allinseattle-of-seattle-area-leaders-raises-27m.html.

22. "We should be scrutinized. I think all large institutions should be scrutinized and examined. It's reasonable." Alyssa Pagano, "Jeff Bezos on Regulating Giant Tech Companies: 'I Expect Us to Be Scrutinized,'" *Business Insider*, November 6, 2018, http:// www.businessinsider.com/jeff-bezos-breaking-up-regulating-amazon-2018-4.

13. The Future of Cities

1. Douglass W. Rae, *City: Urbanism and Its End* (New Haven, CT: Yale University Press, 2003).

2. Seattle City Council Poll, conducted October 22 and 23, 2019, Civic Alliance for a Sound Economy, 2019, https://www.scribd.com/document/432450690/Polling -Memo.

3. In the late 1990s, I served as cochair of former Seattle mayor Paul Schell's losing and winning elections. Paul was an enlightened mayor who was responsible for an award-winning new downtown library, new city hall, rebuilt cultural facilities, and the development of thirty-seven neighborhood plans. He also had to face protests and riots when the World Trade Organization held its global conference in Seattle.

4. Rahm Emanuel, *The Nation City: Why Mayors Are Now Running the World* (New York: Knopf, 2020).

5. Daniel Beckman, "Seattle Police Chief Carmen Best Says City Council's Budget Cuts, Lack of Respect for SPD Drove Her Retirement Decision," *Seattle Times*, August 11, 2020, https://www.seattletimes.com/seattle-news/politics/seattle-police-chief-carmen-best -says-city-council-drove-her-decision-to-abruptly-retire.

Ron Sims, the former county executive called the council's action "reckless": "As a Black man, I found Seattle City Council's treatment of Chief Best to be incredibly offensive. They'll say they're liberals. It didn't give them license to publicly humiliate the first African American Women to be Chief of Police in Seattle's history." Ron Sims, Twitter @simsron, August 10, 2020, https://twitter.com/simsron/status/1293035395207159809.

6. Seattle City Council Poll, October 22 and 23, 2019.

7. Robert McCartney, Jonathan O'Connell, and Patricia Sullivan, "Facing Opposition, Amazon Reconsiders N.Y. Headquarters Site, Two Officials Say," *Washington Post*, February 8, 2019, https://www.washingtonpost.com/local/virginia

-politics/facing-opposition-amazon-reconsiders-ny-headquarters-site-two-officials
-say/2019/02/08/451ffc52-2a19-11e9-b011-d8500644dc98_story.html; and Eli Rosenberg
and Reis Thiebault, "Amazon Had New York City in the Bag. Then Left-Wing
Activists Got Fired Up," *Washington Post*, February 4, 2019, https://www.washingtonpost
.com/nation/2019/02/14/how-amazons-big-plans-new-york-city-were-thwarted-by
-citys-resurgent-left-wing/.

8. Tyler Clifford, " 'This Is a Union Town'—NYC Councilman Says Amazon's HQ2
Is 'Antithetical' to Our Values," CNBC, February 11, 2019, https://www.cnbc.com/2019
/02/11/amazon-hq2-is-antithetical-to-our-union-values-nyc-councilman-says.html.

9. Jim Collins, *Good to Great: Why Some Companies Make the Leap . . . And
Others Don't* (New York: HarperBusiness, 2001), 13.

10. "Automatic Vehicles Tactical Plan," City of Toronto, accessed April 1, 2021,
https://www.toronto.ca/services-payments/streets-parking-transportation/transportation
-projects/automated-vehicles/draft-automated-vehicle-tactical-plan-2019-2021/.

11. "2019 was a groundbreaking year in Columbus' smart city journey. Self-
driving shuttles traveled 19,000 miles on Columbus city streets. Hundreds of electric
vehicle charging stations were brought online." Columbus Partnership Year in Review,
2019, page 17, https://static1.squarespace.com/static/5e73e44054304f65ff5c5d55/t
/5e86924bc46c8d7790464c19/1585877588573/TCP_2019_YearinReview+.pdf.

Index

Page numbers in *italics* indicate figures.